Embedded Systems Design with the
Atmel AVR Microcontroller
Part II

Synthesis Lectures on Digital Circuits and Systems

Editor
Mitchell A. Thornton, *Southern Methodist University*

Pragmatic Circuits: Frequency Domain
William J. Eccles
2006

Pragmatic Circuits: Signals and Filters
William J. Eccles
2006

High-Speed Digital System Design
Justin Davis
2006

Introduction to Logic Synthesis using Verilog HDL
Robert B.Reese, Mitchell A.Thornton
2006

Microcontrollers Fundamentals for Engineers and Scientists
Steven F. Barrett, Daniel J. Pack
2006

Embedded Systems Design with the Atmel AVR Microcontroller – Part II

Steven F. Barrett

ISBN: 978-3-031-79808-5   paperback
ISBN: 978-3-031-79809-2   ebook

DOI 10.1007/978-3-031-79809-2

A Publication in the Springer series
SYNTHESIS LECTURES ON DIGITAL CIRCUITS AND SYSTEMS

Lecture #25
Series ISSN
Synthesis Lectures on Digital Circuits and Systems
Print 1932-3166   Electronic 1932-3174

Embedded Systems Design with the
Atmel AVR Microcontroller
Part II

Steven F. Barrett
University of Wyoming

SYNTHESIS LECTURES ON DIGITAL CIRCUITS AND SYSTEMS #25

ABSTRACT

This textbook provides practicing scientists and engineers an advanced treatment of the Atmel AVR microcontroller. This book is intended as a follow on to a previously published book, titled "Atmel AVR Microcontroller Primer: Programming and Interfacing." Some of the content from this earlier text is retained for completeness. This book will emphasize advanced programming and interfacing skills. We focus on system level design consisting of several interacting microcontroller subsystems. The first chapter discusses the system design process. Our approach is to provide the skills to quickly get up to speed to operate the internationally popular Atmel AVR microcontroller line by developing systems level design skills. We use the Atmel ATmega164 as a representative sample of the AVR line. The knowledge you gain on this microcontroller can be easily translated to every other microcontroller in the AVR line. In succeeding chapters, we cover the main subsystems aboard the microcontroller, providing a short theory section followed by a description of the related microcontroller subsystem with accompanying software for the subsystem. We then provide advanced examples exercising some of the features discussed. In all examples, we use the C programming language. The code provided can be readily adapted to the wide variety of compilers available for the Atmel AVR microcontroller line. We also include a chapter describing how to interface the microcontroller to a wide variety of input and output devices. The book concludes with several detailed system level design examples employing the Atmel AVR microcontroller.

KEYWORDS

Atmel microcontroller, Atmel AVR, ATmega164, microcontroller interfacing, embedded systems design

Contents

Acknowledgments

I would like to dedicate this book to my close friend and writing partner Dr. Daniel Pack, Ph.D., P.E. Daniel elected to "sit this one out" because of a thriving research program in unmanned aerial vehicles (UAVs). Daniel took a very active role in editing the final manuscript of this text. Also, much of the writing is his from earlier Morgan & Claypool projects. In 2000, Daniel suggested that we might write a book together on microcontrollers. I had always wanted to write a book but I thought that's what other people did. With Daniel's encouragement we wrote that first book (and five more since then). Daniel is a good father, good son, good husband, brilliant engineer, a work ethic second to none, and a good friend. To you, good friend, I dedicate this book. I know that we will do many more together.

Steven F. Barrett
October 2009

Preface

In 2006 Morgan & Claypool Publishers (M&C) released the textbook, titled "Microcontrollers Fundamentals for Engineers and Scientists." The purpose of the textbook was to provide practicing scientists and engineers with a tutorial on the fundamental concepts and the use of microcontrollers. The textbook presented the fundamental concepts common to all microcontrollers. This book was followed in 2008 with "Atmel AVR Microcontroller Primer: Programming and Interfacing." The goal for writing this follow-on book was to provide details on a specific microcontroller family – the Atmel AVR Microcontroller. This book is the third in the series. In it the emphasis is on system level design and advanced microcontroller interfacing and programming concepts. Detailed examples are provided throughout the text.

APPROACH OF THE BOOK

We assume the reader is already familiar with the Atmel AVR microcontroller line. If this is not the case, we highly recommend a first read of "Atmel AVR Microcontroller Primer: Programming and Interfacing." Although some of the content from this earlier volume is retained in this current book for completeness, the reader will be much better served with a prior solid background in the Atmel AVR microcontroller family.

Chapter 1 contains an overview of embedded systems level design. Chapter 2 presents a brief review of the Atmel AVR subsystem capabilities and features. Chapters 3 through 7 provide the reader with a detailed treatment of the subsystems aboard the AVR microcontroller. Chapter 8 ties together the entire book with several examples of system level design.

Steven F. Barrett
October 2009

CHAPTER 6

Timing Subsystem

Objectives: After reading this chapter, the reader should be able to

- Explain key timing system related terminology.

- Compute the frequency and the period of a periodic signal using a microcontroller.

- Describe functional components of a microcontroller timer system.

- Describe the procedure to capture incoming signal events.

- Describe the procedure to generate time critical output signals.

- Describe the timing related features of the Atmel ATmega164.

- Describe the four operating modes of the Atmel ATmega164 timing system.

- Describe the register configurations for the ATmega164's Timer 0, Timer 1, and Timer 2.

- Program the ATmega164 timer system for a variety of applications.

6.1 OVERVIEW

One of the most important reasons for using microcontrollers in embedded systems is the capabilities of microcontrollers to perform time related tasks. In a simple application, one can program a microcontroller system to turn on or turn off an external device at a programmed time. In a more involved application, we can use a microcontroller to generate complex digital waveforms with varying pulse widths to control the speed of a DC motor. In this chapter, we review the capabilities of the Atmel ATmega164 microcontroller to perform time related functions. We begin with a review of timing related terminology. We then provide an overview of the general operation of a timing system followed by the timing system features aboard the ATmega164. Next, we present a detailed discussion of each of its timing channels: Timer 0, Timer 1, and Timer 2 and their different modes of operation.

6.2 TIMING RELATED TERMINOLOGY

6.2.1 FREQUENCY

Consider signal $x(t)$ that repeats itself. We call this signal periodic with period T, if it satisfies

$$x(t) = x(t + T).$$

To measure the frequency of a periodic signal, we count the number of times a particular event repeats within a one second period. The unit of frequency is the Hertz or cycles per second. For example, a sinusoidal signal with the 60 Hz frequency means that a full cycle of a sinusoid signal repeats itself 60 times each second or every 16.67 ms.

6.2.2   PERIOD
The reciprocal of frequency is the period of a waveform. If an event occurs with a rate of 1 Hz, the period of that event is 1 second. To find a period, given a frequency of a signal, or vice versa, we simply need to remember their inverse relationship $f = \frac{1}{T}$ where f and T represent a frequency and the corresponding period, respectively. Both periods and frequencies of signals are often used to specify timing constraints of embedded systems. For example, when your car is on a wintery road and slipping, the engineers who designed your car configured the anti-slippage unit to react within some millisecond period, say 20 milliseconds. The constraint then forces the design team that monitors the slippage to program their monitoring system to check a slippage at a rate of 50 Hz.

6.2.3   DUTY CYCLE
In many applications, periodic pulses are used as control signals. A good example is the use of a periodic pulse to control a servo motor. To control the direction and sometimes the speed of a motor, a periodic pulse signal with a changing duty cycle over time is used. The periodic pulse signal shown in Figure 6.1a) is on for 50 percent of the signal period and off for the rest of the period. The pulse shown in (b) is on for only 25 percent of the same period as the signal in (a) and off for 75 percent of the period. The duty cycle is defined as the percentage of one period a signal is on. Therefore, we call the signal in Figure 6.1(a) as a periodic pulse signal with a 50 percent duty cycle and the corresponding signal in (b), a periodic pulse signal with a 25 percent duty cycle.

6.3   TIMING SYSTEM OVERVIEW

The heart of the timing system is the time base. The time base's frequency of a microcontroller is used to generate a baseline clock signal. For a timer system, the system clock is used to update the contents of a special register called a free running counter. The job of a free running counter is to count up (increment) each time it sees a rising edge (or a falling edge) of a clock signal. Thus, if a clock is running at the rate of 2 MHz, the free running counter will count up each 0.5 microseconds. All other timer related units reference the contents of the free running counter to perform input and output time related activities: measurement of time periods, capture of timing events, and generation of time related signals.

The ATmega164 may be clocked internally using a user-selectable resistor capacitor (RC) time base or it may be clocked externally. The RC internal time base is selected using programmable

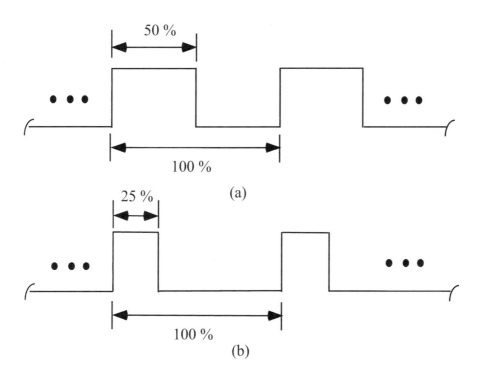

Figure 6.1: Two signals with the same period but different duty cycles. Frame (a) shows a periodic signal with a 50% duty cycle and frame (b) displays a periodic signal with a 25% duty cycle.

fuse bits. We will discuss how to do this in the application section of this chapter. You may choose an internal fixed clock operating frequency of 1, 2, 4 or 8 MHz.

To provide for a wider range of frequency selections an external time source may be used. The external time sources, in order of increasing accuracy and stability, are an external RC network, a ceramic resonator, and a crystal oscillator. The system designer chooses the time base frequency and clock source device appropriate for the application at hand. As previously mentioned, the maximum operating frequency of the ATmega164P with a 5 VDC supply voltage is 20 MHz.

For input time related activities, all microcontrollers typically have timer hardware components that detect signal logic changes on one or more input pins. Such components rely on a free running counter to capture external event times. We can use such ability to measure the period of an incoming signal, the width of a pulse, and the time of a signal logic change.

For output timer functions, a microcontroller uses a comparator, a free running counter, logic switches, and special purpose registers to generate time related signals on one or more output pins. A comparator checks the value of the free running counter for a match with the contents of another special purpose register where a programmer stores a specified time in terms of the free running counter value. The checking process is executed at each clock cycle and when a match occurs, the corresponding hardware system induces a programmed logic change on a programmed output port pin. Using such capability, one can generate a simple logic change at a designated time incident, a pulse with a desired time width, or a pulse width modulated signal to control servo or Direct Current (DC) motors.

You can also use the timer input system to measure the pulse width of an aperiodic signal. For example, suppose that the times for the rising edge and the falling edge of an incoming signal are 1.5 sec and 1.6 sec, respectively. We can use these values to easily compute the pulse width of 0.1 second.

The second overall goal of the timer system is to generate signals to control external devices. Again, an event simply means a change of logic states on an output pin of a microcontroller at a specified time. Now consider Figure 6.2. Suppose an external device connected to the microcontroller requires a pulse signal to turn itself on. Suppose the particular pulse the external device needs is 2 millisecond wide. In such situations, we can use the free running counter value to synchronize the time of desired logic state changes. Naturally, extending the same capability, we can also generate a periodic pulse with a fixed duty cycle or a varying duty cycle.

From the examples, we discussed above, you may have wondered how a microcontroller can be used to compute absolute times from the relative free running counter values, say 1.5 second and 1.6 second. The simple answer is that we can not do so directly. A programmer must use the relative system clock values and derive the absolute time values. Suppose your microcontroller is clocked by a 2 MHz signal and the system clock uses a 16-bit free running counter. For such a system, each clock period represents 0.5 microsecond and it takes approximately 32.78 milliseconds to count from 0 to 2^{16} (65,536). The timer input system then uses the clock values to compute frequencies, periods, and pulse widths. For example, suppose you want to measure a pulse width of an incoming aperiodic signal. If the rising edge and the falling edge occurred at count values $0010 and $0114, [1] can you find the pulse width when the free running counter is counting at 2 MHz? Let's first convert the two values into their corresponding decimal values, 276 and 16. The pulse width of the signal in the number of counter value is 260. Since we already know how long it takes for the system to count one, we can readily compute the pulse width as $260 \times 0.5 \, microseconds \, = \, 130 \, microseconds$.

Our calculations do not take into account time increments lasting longer than the rollover time of the counter. When a counter rolls over from its maximum value back to zero, a flag is set to notify the processor of this event. The rollover events may be counted to correctly determine the overall elapsed time of an event.

Elapsed time may be calculated using:

$$elapsed \; clock \; ticks \; = \; (n \; \times \; 2^b) \; + \; (stop \; count \; - \; start \; count) \; [clock \; ticks]$$

[1]The $ symbol represents that the following value is in a hexadecimal form.

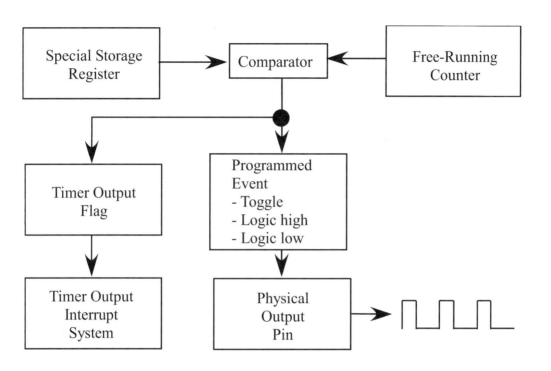

Figure 6.2: A diagram of a timer output system.

$$elapsed\ time = (elapsed\ clock\ ticks) \times (FRC\ clock\ period)\ [seconds]$$

In this first equation, "n" is the number of Timer Overflow Flag (TOF) events, that occur between the start and stop events, and "b" is the number of bits in the timer counter. The equation yields the elapsed time in clock ticks. To convert to seconds the number of clock ticks are multiplied by the period of the clock source of the free running counter.

6.4   APPLICATIONS

In this section, we consider some important uses of the timer system of a microcontroller to (1) measure an input signal timing event, termed input capture, (2) to count the number of external signal occurrences, (3) to generate timed signals—termed output compare, and, finally, (4) to generate pulse width modulated signals. We first start with a case of measuring the time duration of an incoming signal.

6.4.1   INPUT CAPTURE—MEASURING EXTERNAL TIMING EVENT

In many applications, we are interested in measuring the elapsed time or the frequency of an external event using a microcontroller. Using the hardware and functional units discussed in the previous sections, we now present a procedure to accomplish the task of computing the frequency of an incoming periodic signal. Figure 6.3 shows an incoming periodic signal to our microcontroller.

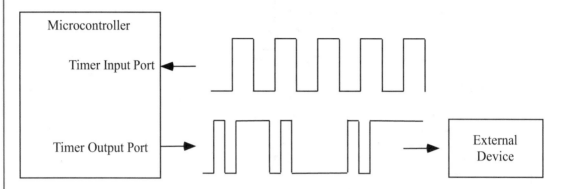

Figure 6.3: Use of the timer input and output systems of a microcontroller. The signal on top is fed into a timer input port. The captured signal is subsequently used to compute the input signal frequency. The signal on the bottom is generated using the timer output system. The signal is used to control an external device.

The first necessary step for the current task is to turn on the timer system. To reduce power consumption, a microcontroller usually does not turn on all of its functional systems after reset until they are needed. In addition to a separate timer module, many microcontroller manufacturers allow a programmer to choose the rate of a separate timer clock that governs the overall functions of a timer module.

Once the timer is turned on and the clock rate is selected, a programmer must configure the physical port to which the incoming signal arrives. This step is done using a special input timer port configuration register. The next step is to program the input event to capture. In our current example, we should capture two consecutive rising edges or falling edges of the incoming signal. Again, the programming portion is done by storing an appropriate setup value to a special register.

Now that the input timer system is configured appropriately, you now have two options to accomplish the task. The first one is the use of a polling technique; the microcontroller continuously polls a flag, which holds a logic high signal when a programmed event occurs on the physical pin. Once the microcontroller detects the flag, it needs to clear the flag and record the time when the flag was set using another special register that captures the time of the associated free running counter value. The program needs to continue to wait for the next flag which indicates the end of one period of the incoming signal. A programmer then needs to record the newly acquired captured time represented in the form of a free running counter value again. The period of the signal can now be computed by computing the time difference between the two captured event times, and, based on the clock speed of the microcontroller, the programmer can compute the actual time changes and consequently the frequency of the signal.

In many cases, a microcontroller can't afford the time to poll for one event. Such situation introduces the second method: interrupt systems. Most microcontroller manufacturers have developed built-in interrupt systems with their timer input modules. Instead of continuously polling for a flag, a microcontroller performs other tasks and relies on its interrupt system to detect the programmed event. The task of computing the period and the frequency is the same as the first method, except that the microcontroller will not be tied down to constantly checking the flag, increasing the efficient use of the microcontroller resources. To use interrupt systems, of course, we must pay the price by appropriately configuring the interrupt systems to be triggered when a desired event is detected. Typically, additional registers must be configured, and a special program called an interrupt service routine must be written.

Suppose that for an input capture scenario the two captured times for the two rising edges are $1000 and $5000, respectively. Note that these values are not absolute times but the representations of times reflected as the values of the free running counter. The period of the signal is $4000 or 16384 in a decimal form. If we assume that the timer clock runs at 10 MHz, the period of the signal is 1.6384 msec, and the corresponding frequency of the signal is approximately 610.35 Hz.

6.4.2   COUNTING EVENTS

The same capability of measuring the period of a signal can also be used to simply count external events. Suppose we want to count the number of logic state changes of an incoming signal for a given period of time. Again, we can use the polling technique or the interrupt technique to accomplish the task. For both techniques, the initial steps of turning on a timer and configuring a physical input port pin are the same. In this application, however, the programmed event should be any logic state changes instead of looking for a rising or a falling edge as we have done in the previous section. If the polling technique is used, at each event detection, the corresponding flag must be cleared and a counter must be updated. If the interrupt technique is used, one must write an interrupt service routine within which the flag is cleared and a counter is updated.

6.4.3   OUTPUT COMPARE—GENERATING TIMING SIGNALS TO INTER-FACE EXTERNAL DEVICES

In the previous two sections, we considered two applications of capturing external incoming signals. In this subsection and the next one, we consider how a microcontroller can generate time critical signals for external devices. Suppose in this application, we want to send a signal shown in Figure 6.3 to turn on an external device. The timing signal is arbitrary but the application will show that a timer output system can generate any desired time related signals permitted under the timer clock speed limit of the microcontroller.

Similar to the use of the timer input system, one must first turn on the timer system and configure a physical pin as a timer output pin using special registers. In addition, one also needs to program the desired external event using another special register associated with the timer output system. To generate the signal shown in Figure 6.3, one must compute the time required between the rising and the falling edges. Suppose that the external device requires a pulse which is 2 milliseconds wide to be activated. To generate the desired pulse, one must first program the logic state for the particular pin to be low and set the time value using a special register with respect to the contents of the free running counter. As was mentioned in Section 5.2, at each clock cycle, the special register contents are compared with the contents of the free running counter and when a match occurs, the programmed logic state appears on the designated hardware pin. Once the rising edge is generated, the program then must reconfigure the event to be a falling edge (logic state low) and change the contents of the special register to be compared with the free running counter. For the particular example in Figure 6.3, let's assume that the main clock runs at 2 MHz, the free running counter is a 16 bit counter, and the name of the special register (16 bit register) where we can put appropriate values is output timer register. To generate the desired pulse, we can put $0000 first to the output timer register, and after the rising edge has been generated, we need to change the program event to a falling edge and put $0FA0 or 4000 in decimal to the output timer register. As was the case with the input timer system module, we can use output timer system interrupts to generate the desired signals as well.

6.4.4   INDUSTRIAL IMPLEMENTATION CASE STUDY (PWM)

In this section, we discuss a well-known method to control the speed of a DC motor using a pulse width modulated (PWM) signal. The underlying concept is as follows. If we turn on a DC motor and provide the required voltage, the motor will run at its maximum speed. Suppose we turn the motor on and off rapidly, by applying a periodic signal. The motor at some point can not react fast enough to the changes of the voltage values and will run at the speed proportional to the average time the motor was turned on. By changing the duty cycle, we can control the speed of a DC motor as we desire. Suppose again, we want to generate a speed profile shown in Figure 6.4. As shown in the figure, we want to accelerate the speed, maintain the speed, and decelerate the speed for a fixed amount of time.

The first task necessary is again to turn on the timer system, configure a physical port, and program the event to be a rising edge. As a part of the initialization process, we need to put $0000 to the output timer register we discussed in the previous subsection. Once the rising edge is generated, the program then needs to modify the event to a falling edge and change the contents of the output timer register to a value proportional to a desired duty cycle. For example, if we want to start off with 25% duty cycle, we need to input $4000 to the register, provided that we are using a 16 bit free running counter. Once the falling edge is generated, we now need to go back and change the event to be a rising edge and the contents of the output timer counter value back to $0000. If we want

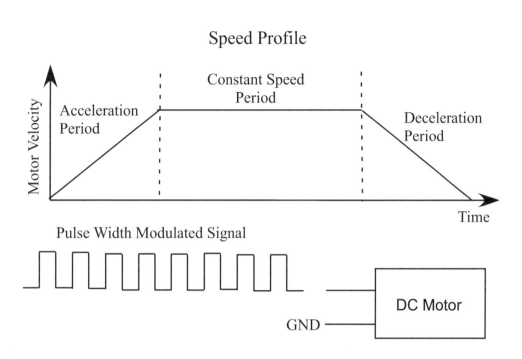

Figure 6.4: The figure shows the speed profile of a DC motor over time when a pulse-width-modulated signal is applied to the motor.

to continue to generate a 25% duty cycle signal, then we must repeat the process indefinitely. Note that we are using the time for a free running counter to count from $0000 to $FFFF as one period.

Now suppose we want to increase the duty cycle to 50% over 1 sec and that the clock is running at 2 MHz. This means that the free running counter counts from $0000 to $FFFF every 32.768-milliseconds, and the free running counter will count from $0000 to $FFFF, approximately 30.51 times over the period of one second. That is, we need to increase the pulse width from $4000 to $8000 in approximately 30 turns, or approximately 546 clock counts every turn.

6.5   OVERVIEW OF THE ATMEL TIMERS

The Atmel ATmega164 is equipped with a flexible and powerful three channel timing system. The timer channels are designated Timer 0, Timer 1, and Timer 2. In this section, we review the operation of the timing system in detail. We begin with an overview of the timing system features followed by a detailed discussion of timer channel 0. Space does not permit a complete discussion of the other two timing channels; we review their complement of registers and highlight their features not contained in our discussion of timer channel 0. The information provided on timer channel 0 is readily adapted to the other two channels.

The features of the timing system are summarized in Figure 6.5. Timer 0 and 2 are 8-bit timers; whereas, Timer 1 is a 16-bit timer. Each timing channel is equipped with a prescaler. The prescaler is used to subdivide the main microcontroller clock source (designated $f_{clk\_I/O}$ in upcoming diagrams) down to the clock source for the timing system (clk_{Tn}).

Timer 0	Timer 1	Timer 2
- 8-bit timer/counter	- 16-bit timer/counter	- 8-bit timer/counter
- 10-bit clock prescaler	- 10-bit clock prescaler	- 10-bit clock prescaler
- Functions:	- Functions:	- Functions:
-- Pulse width modulation	-- Pulse width modulation	-- Pulse width modulation
-- Frequency generation	-- Frequency generation	-- Frequency generation
-- Event counter	-- Event counter	-- Event counter
-- Output compare -- 2 ch	-- Output compare -- 2 ch	-- Output compare -- 2 ch
- Modes of operation:	-- Input capture	- Modes of operation:
-- Normal	- Modes of operation:	-- Normal
-- Clear timer on	-- Normal	-- Clear timer on
compare match (CTC)	-- Clear timer on	compare match (CTC)
-- Fast PWM	compare match (CTC)	-- Fast PWM
-- Phase correct PWM	-- Fast PWM	-- Phase correct PWM
	-- Phase correct PWM	

Figure 6.5: Atmel timer system overview.

Each timing channel has the capability to generate pulse width modulated signals, generate a periodic signal with a specific frequency, count events, and generate a precision signal using the output compare channels. Additionally, Timer 1 is equipped with the Input Capture feature.

All of the timing channels may be configured to operate in one of four operational modes designated : Normal, Clear Timer on Compare Match (CTC), Fast PWM, and Phase Correct PWM. We provide more information on these modes shortly.

6.6 TIMER 0 SYSTEM

In this section, we discuss the features, overall architecture, modes of operation, registers, and programming of Timer 0. This information may be readily adapted to Timer 1 and Timer 2.

A Timer 0 block diagram is shown in Figure 6.6. The clock source for Timer 0 is provided via an external clock source at the T0 pin (PB0) of the microcontroller. Timer 0 may also be clocked internally via the microcontroller's main clock ($f_{clk\_I/O}$). This clock frequency may be too rapid for many applications. Therefore, the timing system is equipped with a prescaler to subdivide the main clock frequency down to timer system frequency (clk_{Tn}). The clock source for Timer 0 is selected using the CS0[2:0] bits contained in the Timer/Counter Control Register B (TCCR0B). The TCCR0A register contains the WGM0[1:0] bits and the COM0A[1:0] (and B) bits. Whereas, the TCCR0B register contains the WGM0[2] bit. These bits are used to select the mode of operation for Timer 0 as well as tailor waveform generation for a specific application.

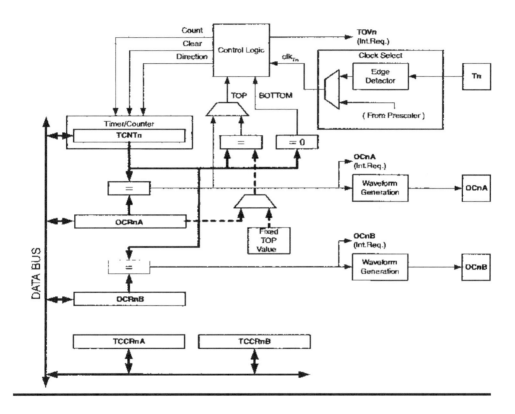

Figure 6.6: Timer 0 block diagram. Figure used with permission Atmel, Inc.

The timer clock source (clk_{Tn}) is fed to the 8-bit Timer/Counter Register (TCNT0). This register is incremented (or decremented) on each clk_{Tn} clock pulse. Timer 0 is also equipped with two 8-bit comparators that constantly compares the numerical content of TCNT0 to the Output Compare Register A (OCR0A) and Output Compare Register B (OCR0B). The compare signal from the 8-bit comparator is fed to the waveform generators. The waveform generators have a number of inputs to perform different operations with the timer system.

The BOTTOM signal for the waveform generation and the control logic, shown in Figure 6.7, is asserted when the timer counter TCNT0 reaches all zeroes (0x00). The MAX signal for the control logic unit is asserted when the counter reaches all ones (0xFF). The TOP signal for the waveform generation is asserted by either reaching the maximum count values of 0xFF on the TCNT0 register or reaching the value set in the Output Compare Register 0 A (OCR0A) or B. The setting for the TOP signal will be determined by the Timer's mode of operation.

Timer 0 also uses certain bits within the Timer/Counter Interrupt Mask Register 0 (TIMSK0) and the Timer/Counter Interrupt Flag Register 0 (TIFR0) to signal interrupt related events.

6.6.1   MODES OF OPERATION

Each of the timer channels may be set for a specific mode of operation: normal, clear timer on compare match (CTC), fast PWM, and phase correct PWM. The system designer chooses the correct mode for the application at hand. The timer modes of operation are summarized in Figure 6.7. A specific mode of operation is selected using the Waveform Generation Mode bits located in Timer/Control Register A (TCCR0A) and Timer/Control Register B (TCCR0B).

6.6.1.1   Normal Mode

In the normal mode, the timer will continually count up from 0x00 (BOTTOM) to 0xFF (TOP). When the TCNT0 returns to zero on each cycle of the counter the Timer/Counter Overflow Flag (TOV0) will be set. The normal mode is useful for generating a periodic "clock tick" that may be used to calculate elapsed real time or provide delays within a system. We provide an example of this application in Section 5.9.

6.6.1.2   Clear Timer on Compare Match (CTC)

In the CTC mode, the TCNT0 timer is reset to zero every time the TCNT0 counter reaches the value set in Output Compare Register A (OCR0A) or B. The Output Compare Flag A (OCF0A) or B is set when this event occurs. The OCF0A or B flag is enabled by asserting the Timer/Counter 0 Output Compare Math Interrupt Enable (OCIE0A) or B flag in the Timer/Counter Interrupt Mask Register 0 (TIMSK0) and when the I-bit in the Status Register is set to one.

The CTC mode is used to generate a precision digital waveform such as a periodic signal or a single pulse. The user must describe the parameters and key features of the waveform in terms of Timer 0 "clock ticks." When a specific key feature is reached within the waveform, the next key feature may be set into the OCR0A or B register.

6.6.1.3   Phase Correct PWM Mode

In the Phase Correct PWM Mode, the TCNT0 register counts from 0x00 to 0xFF and back down to 0x00, continually. Every time the TCNT0 value matches the value set in the OCR0A or B register, the OCF0A or B flag is set and a change in the PWM signal occurs.

6.6.1.4   Fast PWM

The Fast PWM mode is used to generate a precision PWM signal of a desired frequency and duty cycle. It is called the Fast PWM because its maximum frequency is twice that of the Phase Correct

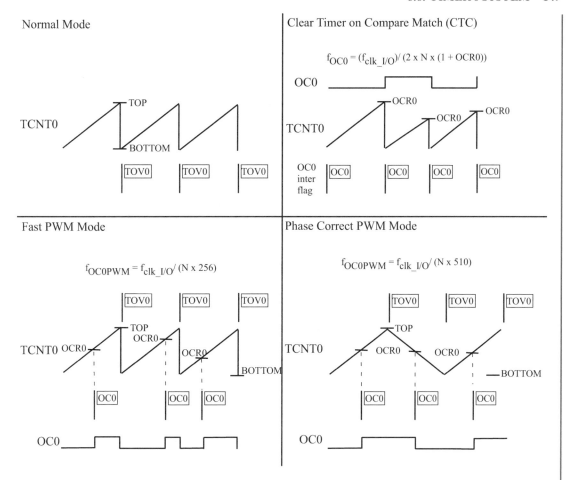

Figure 6.7: Timer 0 modes of operation.

PWM mode. When the TCNT0 register value reaches the value set in the OCR0A or B register, it will cause a change in the PWM output as prescribed by the system designer. It continues to count up to the TOP value at which time the Timer/Counter 0 Overflow Flag is set.

6.6.2   TIMER 0 REGISTERS
A summary of the Timer 0 registers are shown in Figure 6.8.

6.6.2.1 Timer/Counter Control Registers A and B (TCCR0A and TCCR0B)
The TCCR0 register bits are used to:

Timer/Counter Control Register A (TCCR0A)

COM0A1	COM0A0	COM0B1	COM0B0	---	---	WGM01	WGM00

7   0

Timer/Counter Control Register B (TCCR0B)

FOC0A	FOC0B	---	---	WGM02	CS02	CS01	CS00

7   0

Timer/Counter Register (TCNT0)

7   0

Output Compare Register A (OCR0A)

7   0

Output Compare Register B (OCR0B)

7   0

Timer/Counter Interrupt Mask Register 0 (TIMSK0)

---	---	---	---	---	OCIE0B	OCIE0A	TOIE0

7   0

Timer/Counter Interrupt Flag REgister 0 (TIFR0)

---	---	---	---	---	OCF0B	OCF0A	TOV0

7   0

Figure 6.8: Timer 0 registers.

- Select the operational mode of Timer 0 using the Waveform Mode Generation (WGM0[2:0]) bits,

- Determine the operation of the timer within a specific mode with the Compare Match Output Mode (COM0A[1:0] or COM0B[1:0] or) bits, and

- Select the source of the Timer 0 clock using Clock Select (CS0[2:0]) bits.

The bit settings for the TCCR0 register are summarized in Figure 6.9.

6.6.2.2 Timer/Counter Register(TCNT0)

The TCNT0 is the 8-bit counter for Timer 0.

CS0[2:0]	Clock Source
000	None
001	$clk_{I/0}$
010	$clk_{I/0}/8$
011	$clk_{I/0}/64$
100	$clk_{I/0}/8clk_{I/0}/256$
101	$clk_{I/0}/8clk_{I/0}/1024$
110	External clock on T0 (falling edge trigger)
111	External clock on T1 (rising edge trigger)

Clock Select

Timer/Counter Control Register B (TCCR0B)

FOC0A	FOC0B	---	---	WGM02	CS02	CS01	CS00

7   0

Timer/Counter Control Register A (TCCR0A)

COM0A1	COM0A0	COM0B1	COM0B0	---	---	WGM01	WGM00

Waveform Generation Mode

Mode	WGM[02:00]	Mode
0	000	Normal
1	001	PWM, Phase Correct
2	010	CTC
3	011	Fast PWM
4	100	Reserved
5	101	PWM, Phase Correct
6	110	Reserved
7	111	Fast PWM

Compare Output Mode, non-PWM Mode

COM0A[1:0]	Description
00	Normal, OC0A disconnected
01	Toggle OC0A on compare match
10	Clear OC0A on compare match
11	Set OC0A on compare match

Compare Output Mode, non-PWM Mode

COM0B[1:0]	Description
00	Normal, OC0B disconnected
01	Toggle OC0B on compare match
10	Clear OC0B on compare match
11	Set OC0B on compare match

Compare Output Mode, Fast PWM Mode

COM0A[1:0]	Description
00	Normal, OC0A disconnected
01	WGM02 = 0: normal operation, OC0A disconnected WGM02 = 1: Toggle OC0A on compare match
10	Clear OC0A on compare match, set OC0A at Bottom (non-inverting mode)
11	Set OC0A on compare match, clear OC0A at Bottom (inverting mode)

Compare Output Mode, Fast PWM Mode

COM0B[1:0]	Description
00	Normal, OC0B disconnected
01	Reserved
10	Clear OC0B on compare match, set OC0B at Bottom (non-inverting mode)
11	Set OC0B on compare match, clear OC0B at Bottom (inverting mode)

Compare Output Mode, Phase Correct PWM

COM0A[1:0]	Description
00	Normal, OC0A disconnected
01	WGM02 = 0: normal operation, OC0A disconnected WGM02 = 1: Toggle OC0A on compare match
10	Clear OC0A on compare match, when upcounting. Set OC0A on compare match when down counting
11	Set OC0A on compare match, when upcounting. Set OC0A on compare match when down counting

Compare Output Mode, Phase Correct PWM

COM0B[1:0]	Description
00	Normal, OC0B disconnected
01	Reserved
10	Clear OC0B on compare match, when upcounting. Set OC0B on compare match when down counting
11	Set OC0B on compare match, when upcounting. Set OC0B on compare match when down counting

Figure 6.9: Timer/Counter Control Registers A and B (TCCR0A and TCCR0B) bit settings.

6.6.2.3 Output Compare Registers A and B (OCR0A and OCR0B)

The OCR0A and B registers holds a user-defined 8-bit value that is continuously compared to the TCNT0 register.

6.6.2.4 Timer/Counter Interrupt Mask Register (TIMSK0)

Timer 0 uses the Timer/Counter 0 Output Compare Match Interrupt Enable A and B (OCIE0A and B) bits and the Timer/Counter 0 Overflow Interrupt Enable (TOIE0) bit. When the OCIE0A or B bit and the I-bit in the Status Register are both set to one, the Timer/Counter 0 Compare Match interrupt is enabled. When the TOIE0 bit and the I-bit in the Status Register are both set to one, the Timer/Counter 0 Overflow interrupt is enabled.

6.6.2.5 Timer/Counter Interrupt Flag Register 0 (TIFR0)

Timer 0 uses the Output Compare Flag A or B (OCF0A and OCF0B) which sets for an output compare match. Timer 0 also uses the Timer/Counter 0 Overflow Flag (TOV0) which sets when Timer/Counter 0 overflows.

6.7 TIMER 1

Timer 1 is a 16-bit timer/counter. It shares many of the same features of the Timer 0 channel. Due to limited space the shared information will not be repeated. Instead, we concentrate on the enhancements of Timer 1 which include an additional output compare channel and also the capability for input capture. The block diagram for Timer 1 is shown in Figure 6.10.

As discussed earlier in the chapter, the input capture feature is used to capture the characteristics of an input signal including period, frequency, duty cycle, or pulse length. This is accomplished by monitoring for a user-specified edge on the ICP1 microcontroller pin. When the desired edge occurs, the value of the Timer/Counter 1 (TCNT1) register is captured and stored in the Input Capture Register 1 (ICR1).

6.7.1 TIMER 1 REGISTERS

The complement of registers supporting Timer 1 are shown in Figure 6.11. Each register will be discussed in turn.

6.7.1.1 TCCR1A and TCCR1B registers

The TCCR1 register bits are used to:

- Select the operational mode of Timer 1 using the Waveform Mode Generation (WGM1[3:0]) bits,

- Determine the operation of the timer within a specific mode with the Compare Match Output Mode (Channel A: COM1A[1:0] and Channel B: COM1B[1:0]) bits, and

- Select the source of the Timer 1 clock using Clock Select (CS1[2:0]) bits.

The bit settings for the TCCR1A and TCCR1B registers are summarized in Figure 6.12.

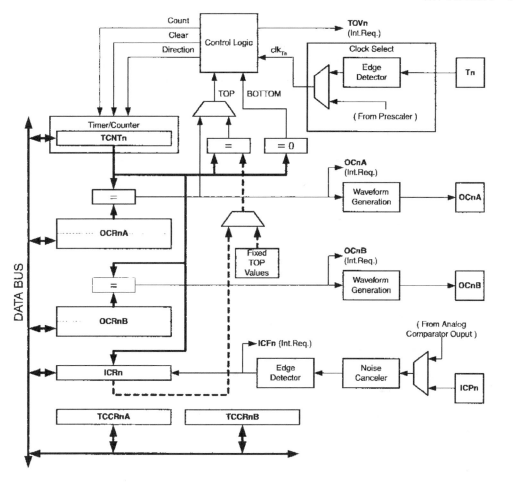

Figure 6.10: Timer 1 block diagram. (Figure used with Permission, Atmel, Inc.)

6.7.1.2   Timer/Counter Register 1 (TCNT1H/TCNT1L)
The TCNT1 is the 16-bit counter for Timer 1.

6.7.1.3   Output Compare Register 1 (OCR1AH/OCR1AL)
The OCR1A register holds a user-defined 16-bit value that is continuously compared to the TCNT1 register when Channel A is used.

6.7.1.4   OCR1BH/OCR1BL
The OCR1B register holds a user-defined 16-bit value that is continuously compared to the TCNT1 register when Channel B is used.

Timer/Counter 1 Control Register A (TCCR1A)

COM1A1	COM1A0	COM1B1	COM1B0	---	---	WGM11	WGM10

7   0

Timer/Counter 1 Control Register B (TCCR1B)

ICNC1	ICES1	---	WGM13	WGM12	CS12	CS11	CS10

7   0

Timer/Counter 1 Control Register C (TCCR1C)

FOC1A	FOC1B	---	---	---	---	---	---

7   0

Timer Counter1 (TCNT1H/TCNT1L)

15							8
7							0

Output Compare Register 1 A (OCR1AH/OCR1AL)

15							8
7							0

Output Compare Register 1 B (OCR1BH/OCR1BL)

15							8
7							0

Input Capture Register 1 (ICR1H/ICR1L)

15							8
7							0

Timer/Counter Interrupt Mask Register 1 (TIMSK1)

---	---	ICIE1	---	---	OCIE1B	OCIE1A	TOIE1

7   0

Timer/Counter 1 Interrupt Flag REgister (TIFR1)

---	---	ICF1	---	---	OCF1B	OCF1A	TOV1

7   0

Figure 6.11: Timer 1 registers.

CS0[2:0]	Clock Source
000	None
001	$clk_{I/0}$
010	$clk_{I/0}/8$
011	$clk_{I/0}/64$
100	$clk_{I/0}/8clk_{I/0}/256$
101	$clk_{I/0}/8clk_{I/0}/1024$
110	External clock on T0 (falling edge trigger)
111	External clock on T1 (rising edge trigger)

Clock Select

Timer/Counter 1 Control Register B (TCCR1B)

ICNC1	ICES1	---	WGM13	WGM12	CS12	CS11	CS10
7							0

Timer/Counter 1 Control Register A (TCCR1A)

COM1A1	COM1A0	COM1B1	COM1B0	---	---	WGM11	WGM10
7							0

Waveform Generation Mode

Mode	WGM[13:12:11:10]	Mode
0	0000	Normal
1	0001	PWM, Phase Correct, 8-bit
2	0010	PWM, Phase Correct, 9-bit
3	0011	PWM, Phase Correct, 10-bit
4	0100	CTC
5	0101	Fast PWM, 8-bit
6	0110	Fast PWM, 9-bit
7	0111	Fast PWM, 10-bit
8	1000	PWM, Phase & Freq Correct
9	1001	PWM, Phase & Freq Correct
10	1010	PWM, Phase Correct
11	1011	PWM, Phase Correct
12	1100	CTC
13	1101	Reserved
14	1110	Fast PWM
15	1111	Fast PWM

Normal, CTC

COMx[1:0]	Description
00	Normal, OC1A/1B disconnected
01	Toggle OC1A/1B on compare match
10	Clear OC1A/1B on compare match
11	Set OC1A/1B on compare match

PWM, Phase Correct, Phase & Freq Correct

COMx[1:0]	Description
00	Normal, OC0 disconnected
01	WGM1[3:0] = 9 or 14: toggle OCnA on compare match, OCnB disconnected WGM1[3:0]= other settings, OC1A/1B disconnected
10	Clear OC0 on compare match when up-counting. Set OC0 on compare match when down counting
11	Set OC0 on compare match when up-counting. Clear OC0 on compare match when down counting.

Fast PWM

COMx[1:0]	Description
00	Normal, OC1A/1B disconnected
01	WGM1[3:0] = 9 or 11, toggle OC1A on compare match OC1B disconnected WGM1[3:0] = other settings, OC1A/1B disconnected
10	Clear OC1A/1B on compare match, set OC1A/1B on Compare Match when down counting
11	Set OC1A/1B on compare match when upcounting. Clear OC1A/1B on Compare Match when upcounting

Figure 6.12: TCCR1A and TCCR1B registers.

6.7.1.5 Input Capture Register 1 (ICR1H/ICR1L)

ICR1 is a 16-bit register used to capture the value of the TCNT1 register when a desired edge on ICP1 pin has occurred.

6.7.1.6 Timer/Counter Interrupt Mask Register 1 (TIMSK1)

Timer 1 uses the Timer/Counter 1 Output Compare Match Interrupt Enable (OCIE1A/1B) bits, the Timer/Counter 1 Overflow Interrupt Enable (TOIE1) bit, and the Timer/Counter 1 Input Capture Interrupt Enable (IC1E1) bit. When the OCIE1A/B bit and the I-bit in the Status Register are both set to one, the Timer/Counter 1 Compare Match interrupt is enabled. When the OIE1 bit and the I-bit in the Status Register are both set to one, the Timer/Counter 1 Overflow interrupt is enabled. When the IC1E1 bit and the I-bit in the Status Register are both set to one, the Timer/Counter 1 Input Capture interrupt is enabled.

6.7.1.7 Timer/Counter Interrupt Flag Register (TIFR1)

Timer 1 uses the Output Compare Flag 1 A/B (OCF1A/B) which sets for an output compare A/B match. Timer 1 also uses the Timer/Counter 1 Overflow Flag (TOV1) which sets when Timer/Counter 1 overflows. Timer Channel 1 also uses the Timer/Counter 1 Input Capture Flag (ICF1) which sets for an input capture event.

6.8 TIMER 2

Timer 2 is another 8-bit timer channel similar to Timer 0. The Timer 2 channel block diagram is provided in Figure 6.13. Its registers are summarized in Figure 6.14.

6.8.0.8 Timer/Counter Control Register A and B (TCCR2A and B)

The TCCR2A and B register bits are used to:

- Select the operational mode of Timer 2 using the Waveform Mode Generation (WGM2[2:0]) bits,

- Determine the operation of the timer within a specific mode with the Compare Match Output Mode (COM2A[1:0] and B) bits, and

- Select the source of the Timer 2 clock using Clock Select (CS2[2:0]) bits.

    The bit settings for the TCCR2A and B registers are summarized in Figure 6.15.

6.8.0.9 Timer/Counter Register(TCNT2)

The TCNT2 is the 8-bit counter for Timer 2.

6.8.0.10 Output Compare Register A and B (OCR2A and B)

The OCR2A and B registers hold a user-defined 8-bit value that is continuously compared to the TCNT2 register.

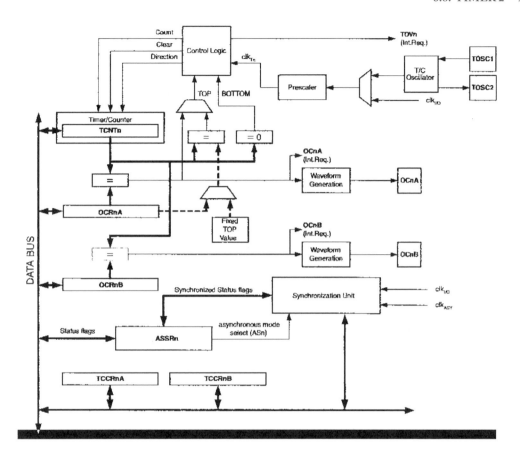

Figure 6.13: Timer 2 block diagram. (Figure used with Permission, Atmel, Inc.)

6.8.0.11 Timer/Counter Interrupt Mask Register 2 (TIMSK2)

Timer 2 uses the Timer/Counter 2 Output Compare Match Interrupt Enable A and B (OCIE2A and B) bits and the Timer/Counter 2 Overflow Interrupt Enable A and B (OIE2A and B) bits. When the OCIE2A or B bit and the I-bit in the Status Register are both set to one, the Timer/Counter 2 Compare Match interrupt is enabled. When the TOIE2 bit and the I-bit in the Status Register are both set to one, the Timer/Counter 2 Overflow interrupt is enabled.

6.8.0.12 Timer/Counter Interrupt Flag Register 2 (TIFR2)

Timer 2 uses the Output Compare Flags 2 A and B (OCF2A and B) which sets for an output compare match. Timer 2 also uses the Timer/Counter 2 Overflow Flag (TOV2) which sets when Timer/Counter 2 overflows.

Timer/Counter Control Register A (TCCR2A)

COM2A1	COM2A0	COM2B1	COM2B0	---	---	WGM21	WGM20

7   0

Timer/Counter Control Register B (TCCR2B)

FOC2A	FOC2B	---	---	WGM22	CS22	CS21	CS20

7   0

Timer/Counter Register (TCNT2)

7   0

Output Compare Register A (OCR2A)

7   0

Output Compare Register B (OCR2B)

7   0

Timer/Counter 2 Interrupt Mask Register (TIMSK2)

---	---	---	---	---	OCIE2B	OCIE2A	TOIE2

7   0

Timer/Counter 2 Interrupt Flag REgister (TIFR2)

---	---	---	---	---	OCF2B	OCF2A	TOV2

7   0

Figure 6.14: Timer 2 registers.

6.9   PROGRAMMING THE TIMER SYSTEM

In this section, we provide several representative examples of using the timer system for various applications. We will provide examples of using the timer system to generate a prescribed delay, to generate a PWM signal, and to capture an input event.

6.9.1   PRECISION DELAY

In this example, we program the ATmega164 to provide a delay of some number of 6.55 ms interrupts. The Timer 0 overflow is configured to occur every 6.55 ms. The overflow flag is used as a "clock tick" to generate a precision delay. To create the delay, the microcontroller is placed in a while loop waiting for the prescribed number of Timer 0 overflows to occur.

CS2[2:0]	Clock Source
000	None
001	$\mathrm{clk_{I/0}}$
010	$\mathrm{clk_{I/0}}/8$
011	$\mathrm{clk_{I/0}}/32$
100	$\mathrm{clk_{I/0}}/64$
101	$\mathrm{clk_{I/0}}/128$
110	$\mathrm{clk_{I/0}}/256$
111	$\mathrm{clk_{I/0}}/1024$

Clock Select

Timer/Counter Control Register B (TCCR2B)

FOC2A	FOC2B	---	---	WGM22	CS22	CS21	CS20

7   0

Timer/Counter Control Register A (TCCR2A)

COM2A1	COM2A0	COM2B1	COM2B0	---	---	WGM21	WGM20

Waveform Generation Mode

Mode	WGM[02:00]	Mode
0	000	Normal
1	001	PWM, Phase Correct
2	010	CTC
3	011	Fast PWM
4	100	Reserved
5	101	PWM, Phase Correct
6	110	Reserved
7	111	Fast PWM

Compare Output Mode, non-PWM Mode

COM2A[1:0]	Description
00	Normal, OC2A disconnected
01	Toggle OC2A on compare match
10	Clear OC2A on compare match
11	Set OC2A on compare match

Compare Output Mode, non-PWM Mode

COM2B[1:0]	Description
00	Normal, OC2B disconnected
01	Toggle OC2B on compare match
10	Clear OC2B on compare match
11	Set OC2B on compare match

Compare Output Mode, Fast PWM Mode

COM2A[1:0]	Description
00	Normal, OC2A disconnected
01	WGM22 = 0: normal operation, OC2A disconnected WGM22 = 1: Toggle OC2A on compare match
10	Clear OC2A on compare match, set OC2A at Bottom (non-inverting mode)
11	Set OC2A on compare match, clear OC2A at Bottom (inverting mode)

Compare Output Mode, Fast PWM Mode

COM2B[1:0]	Description
00	Normal, OC2B disconnected
01	Reserved
10	Clear OC2B on compare match, set OC2B at Bottom (non-inverting mode)
11	Set OC2B on compare match, clear OC2B at Bottom (inverting mode)

Compare Output Mode, Phase Correct PWM

COM2A[1:0]	Description
00	Normal, OC2A disconnected
01	WGM22 = 0: normal operation, OC2A disconnected WGM22 = 1: Toggle OC2A on compare match
10	Clear OC2A on compare match, when upcounting. Set OC2A on compare match when down counting
11	Set OC2A on compare match, when upcounting. Set OC2A on compare match when down counting

Compare Output Mode, Phase Correct PWM

COM2B[1:0]	Description
00	Normal, OC2B disconnected
01	Reserved
10	Clear OC2B on compare match, when upcounting. Set OC2B on compare match when down counting
11	Set OC2B on compare match, when upcounting. Set OC2B on compare match when down counting

Figure 6.15: Timer/Counter Control Register A and B (TCCR2A and B) bit settings.

```
//Function prototypes
void delay(unsigned int number_of_6_55ms_interrupts);
void init_timer0_ovf_interrupt(void);
void timer0_interrupt_isr(void);


                                      //interrupt handler definition
#pragma interrupt_handler timer0_interrupt_isr:19

//door profile data


//***************************************************************************
//int_timer0_ovf_interrupt(): The Timer0 overflow interrupt is being
//employed as a time base for a master timer for this project.  The ceramic
//resonator operating at 10 MHz is divided by 256.  The 8-bit Timer0
//register (TCNT0) overflows every 256 counts or every 6.55 ms.
//***************************************************************************

void init_timer0_ovf_interrupt(void)
{
TCCR0B = 0x04; //divide timer0 timebase
by 256, overflow occurs every 6.55ms
TIMSK0 = 0x01; //enable timer0 overflow interrupt
asm("SEI");    //enable global interrupt
}

//***************************************************************************
//***************************************************************************
//timer0_interrupt_isr:
//Note: Timer overflow 0 is cleared by hardware when executing the
//corresponding interrupt handling vector.
//***************************************************************************


void timer0_interrupt_isr(void)
{
input_delay++;     //input delay processing
```

```
}


//***************************************************************************
//***************************************************************************
//delay(unsigned int num_of_6_55ms_interrupts): this generic delay function
//provides the specified delay as the number of 6.55 ms "clock ticks" from
//the Timer0 interrupt.
//Note: this function is only valid when using a 10 MHz crystal or ceramic
//       resonator
//***************************************************************************

void delay(unsigned int number_of_6_55ms_interrupts)
{
TCNT0 = 0x00;                          //reset timer0
input_delay = 0;
while(input_delay <= number_of_6_55ms_interrupts)
  {
  ;
  }
}


//***************************************************************************
```

6.9.2   PULSE WIDTH MODULATION

The function provided below is used to configure output compare channel B to generate a pulse width modulated signal. An analog voltage provided to ADC Channel 3 is used to set the desired duty cycle from 50 to 100 percent. Note how the PWM ramps up from 0 to the desired speed.

```
//Function Prototypes
void PWM(unsigned int PWM_incr)
{
unsigned int  Open_Speed_int;
float         Open_Speed_float;
int           gate_position_int;


PWM_duty_cycle = 0;
InitADC();                                  //Initialize ADC

   //Read "Open Speed" volt setting PA3
```

```
Open_Speed_int = ReadADC(0x03);
   //Open Speed Setting unsigned int

   //Convert to max duty cycle setting

   //0 VDC = 50% = 127, 5 VDC = 100% =255
Open_Speed_float = ((float)(Open_Speed_int)/(float)(0x0400));

   //Convert volt to PWM constant 127-255
Open_Speed_int = (unsigned int)((Open_Speed_float * 127) + 128.0);
                                       //Configure PWM clock
TCCR1A = 0xA1;
   //freq = resonator/510 = 10 MHz/510

                                       //freq = 19.607 kHz
TCCR1B = 0x01;                         //no clock source division

   //Initiate PWM duty cycle variables
PWM_duty_cycle = 0;

OCR1BH = 0x00;
OCR1BL = (unsigned char)(PWM_duty_cycle);//Set PWM duty cycle CH B to 0%
                                       //Ramp up to Open Speed in 1.6s
OCR1BL = (unsigned char)(PWM_duty_cycle);//Set PWM duty cycle CH B

while (PWM_duty_cycle < Open_Speed_int)
  {
  if(PWM_duty_cycle < Open_Speed_int)    //Increment duty cycle
    PWM_duty_cycle=PWM_duty_cycle + PWM_open_incr;
  OCR1BL = (unsigned char)(PWM_duty_cycle);//Set PWM duty cycle CH B
  }

//Gate continues to open at specified upper speed (PA3)
 :
 :
 :

//*************************************************************************
```

6.9.3   INPUT CAPTURE MODE

This example was developed by Julie Sandberg, BSEE and Kari Fuller, BSEE at the University of Wyoming as part of their senior design project. In this example, the input capture channel is being used to monitor the heart rate (typically 50-120 beats per minute) of a patient. The microcontroller is set to operate at an internal clock frequency of 1 MHz.

```
//*************************************************************************
//initialize_ICP_interrupt: Initialize Timer/Counter 1 for input capture
//*************************************************************************

void initialize_ICP_interrupt(void)
{
TIMSK=0x20;                         //Allows input capture interrupts
SFIOR=0x04;                         //Internal pull-ups disabled
TCCR1A=0x00;                        //No
output comp or waveform generation mode
TCCR1B=0x45;                        //Capture
on rising edge, clock prescalar=1024
TCNT1H=0x00;                        //Initially clear timer/counter 1
TCNT1L=0x00;
asm("SEI");                         //Enable global interrupts
}


//*************************************************************************

void Input_Capture_ISR(void)
{
if(first_edge==0)
  {
  ICR1L=0x00;                       //Clear ICR1 and TCNT1 on first edge
  ICR1H=0x00;
  TCNT1L=0x00;
  TCNT1H=0x00;
  first_edge=1;
  }

else
  {
  ICR1L=TCNT1L;                     //Capture time from TCNT1
  ICR1H=TCNT1H;
```

```c
   TCNT1L=0x00;
   TCNT1H=0x00;
   first_edge=0;
   }

heart_rate();                       //Calculate the heart rate
TIFR=0x20;                          //Clear the input capture flag
asm("RETI");                        //Resets
the I flag to allow global interrupts
}

//*********************************************************************

void heart_rate(void)
{
if(first_edge==0)
  {
  time_pulses_low = ICR1L;          //Read 8 low bits first
  time_pulses_high = ((unsigned int)(ICR1H << 8));
  time_pulses = time_pulses_low | time_pulses_high;
  if(time_pulses!=0)                //1 counter increment = 1.024 ms
    {                               //Divide by 977 to get seconds/pulse
     HR=60/(time_pulses/977);       //(secs/min)/(secs/beat) =bpm
     }
  else
     {

     HR=0;
     }
  }
else
  {
  HR=0;
  }
}

//*********************************************************************
```

6.10   SERVO MOTOR CONTROL WITH THE PWM SYSTEM

A servo motor provides an angular displacement from 0 to 180 degrees. Most servo motors provide the angular displacement relative to the pulse length of repetitive pulses sent to the motor as shown in Figure 6.16. A 1 ms pulse provides an angular displacement of 0 degrees while a 2 ms pulse provides a displacement of 180 degrees. Pulse lengths in between these two extremes provide angular displacements between 0 and 180 degrees. Usually, a 20 to 30 ms low signal is provided between the active pulses.

A test and interface circuit for a servo motor is provided in Figure 6.16. The PB0 and PB1 inputs of the ATmega164 provide for clockwise (CW) and counter-clockwise (CCW) rotation of the servo motor, respectively. The time base for the ATmega164 is set for the 128 KHz internal RC oscillator. Also, the internal time base divide-by-eight circuit is active via a fuse setting. Pulse width modulated signals to rotate the servo motor is provided by the ATmega164. A voltage-follower op amp circuit is used as a buffer between the ATmega164 and the servo motor. Use of an external ceramic resonator at 128 KHz is recommend for this application.

The software to support the test and interface circuit is provided below.

```
//*****************************************************************************
//target controller: ATMEL ATmega324
//
//ATMEL AVR ATmega324PV Controller Pin Assignments
//Chip Port Function I/O Source/Dest Asserted Notes
//PORTB:
//Pin 1 PB0 to active high RC debounced switch - CW
//Pin 2 PB1 to active high RC debounced switch - CCW
//Pin 9 Reset - 1M resistor to Vcc, tact switch to ground, 1.0 uF to ground
//Pin 10 Vcc - 1.0 uF to ground
//Pin 11 Gnd
//Pin 12 ZTT-10.00MT ceramic resonator connection
//Pin 13 ZTT-10.00MT ceramic resonator connection
//Pin 18 PD4 - to servo control input
//Pin 30 AVcc to Vcc
//Pin 31 AGnd to Ground
//Pin 32 ARef to Vcc
//*****************************************************************************

//include files**************************************************************
//ATMEL register definitions for ATmega164
#include<iom164pv.h>
#include<macros.h>
```

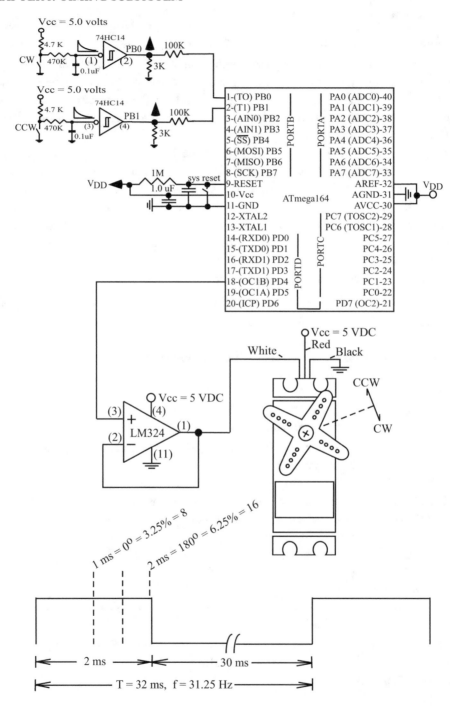

Figure 6.16: Test and interface circuit for a servo motor.

```
//function prototypes**************************************************
void initialize_ports(void);          //initializes ports
void power_on_reset(void);            //returns system to startup state
void read_new_input(void);
 //used to read input change on PORTB
void init_timer0_ovf_interrupt(void);
 //used to initialize timer0 overflow
void InitUSART(void);
void USART_TX(unsigned char data);


//main program*****************************************************
//The main program checks PORTB for user input activity.  If new activity
//is found, the program responds.

//global variables
unsigned char   old_PORTB = 0x08;      //present value of PORTB
unsigned char   new_PORTB;             //new values of PORTB
unsigned int    PWM_duty_cycle;

void main(void)
{
power_on_reset();
 //returns system to startup condition
initialize_ports();
 //return LED configuration to default
InitUSART();
                                       //limited startup features


                                       //internal clock set for 128 KHZ
                                       //fuse set for divide by 8
                                       //configure PWM clock
TCCR1A = 0xA1;
 //freq = oscillator/510 = 128KHz/8/510
                                       //freq = 31.4 Hz
TCCR1B = 0x01;                         //no clock source division

 //duty cycle will vary from 3.1% =
                                       //1 ms = 0 degrees = 8 counts to
```

```c
 //6.2% = 2 ms = 180 degrees = 16 counts

 //initiate PWM duty cycle variables
PWM_duty_cycle = 12;
OCR1BH = 0x00;
OCR1BL = (unsigned char)(PWM_duty_cycle);

//main activity loop - processor will continually cycle through loop for new
//activity. Activity initialized by external signals presented to PORTB[1:0]

while(1)
  {
  _StackCheck();                         //check for stack overflow
  read_new_input();
 //read input status changes on PORTB
  }
}//end main

//Function definitions
//**************************************************************************
//power_on_reset:
//**************************************************************************

void power_on_reset(void)
{
initialize_ports();                      //initialize ports
}


//**************************************************************************
//initialize_ports: provides initial configuration for I/O ports
//**************************************************************************

void initialize_ports(void)
{
//PORTA

DDRA=0xff;                               //PORTA[7:0] output
PORTA=0x00;                              //Turn off pull ups
```

```
//PORTB
DDRB=0xfc;
 //PORTB[7-2] output, PORTB[1:0] input
PORTB=0x00;                              //disable PORTB pull-up resistors

//PORTC
DDRC=0xff;                               //set PORTC[7-0] as output
PORTC=0x00;                              //init low

//PORTD
DDRD=0xff;                               //set PORTD[7-0] as output
PORTD=0x00;                              //initialize low
}

//*****************************************************************************
//*****************************************************************************

//read_new_input: functions polls PORTB for a change in status. If status
//change has occurred, appropriate function for status change is called
//Pin 1 PB0 to active high RC  debounced switch - CW
//Pin 2 PB1 to active high RC debounced switch - CCW
//*****************************************************************************

void read_new_input(void)
{
new_PORTB = (PINB);
if(new_PORTB != old_PORTB){
  switch(new_PORTB){                     //process change in PORTB input
    case 0x01:                  //CW
      while(PINB == 0x01)
        {
        PWM_duty_cycle = PWM_duty_cycle + 1;
        if(PWM_duty_cycle > 16) PWM_duty_cycle = 16;
        OCR1BH = 0x00;
        OCR1BL = (unsigned char)(PWM_duty_cycle);
        }
    break;
```

```
      case 0x02:                              //CCW
        while(PINB == 0x02)
          {
          PWM_duty_cycle = PWM_duty_cycle - 1;
          if(PWM_duty_cycle < 8) PWM_duty_cycle = 8;
          OCR1BH = 0x00;
          OCR1BL = (unsigned char)(PWM_duty_cycle);
          }
        break;


        default:;                             //all other cases
      }                                       //end switch(new_PORTB)
    }                                         //end if new_PORTB
  old_PORTB=new_PORTB;                        //update PORTB
}


//*************************************************************************
```

6.11 PULSE WIDTH MODULATION: AUTOMATED FAN COOLING SYSTEM

In this section, we describe an embedded system application to control the temperature of a room or some device. The system is illustrated in Figure 6.17. An LM34 temperature sensor (PORTA[0]) is used to monitor the instantaneous temperature of the room or device of interest. The current temperature is displayed on the Liquid Crystal Display (LCD).

We send a 1 KHz PWM signal to a cooling fan (M) whose duty cycle is set from 50% to 90% using the potentiometer connected to PORTA[2]. The PWM signal should last until the temperature of the LM34 cools to a value as set by another potentiometer (PORTA[1]). When the temperature of the LM34 falls below the set level, the cooling fan is shut off. If the temperature falls while the fan is active, the PWM signal should gently return to zero, and wait for further temperature changes.

Provided below is the embedded code for the system. This solution was developed by Geoff Luke, UW MSEE, as a laboratory assignment for an Industrial Control class.

```
//*************************************************************************
//Geoff Luke
//EE 5880 - Industrial Controls
//PWM Fan Control
```

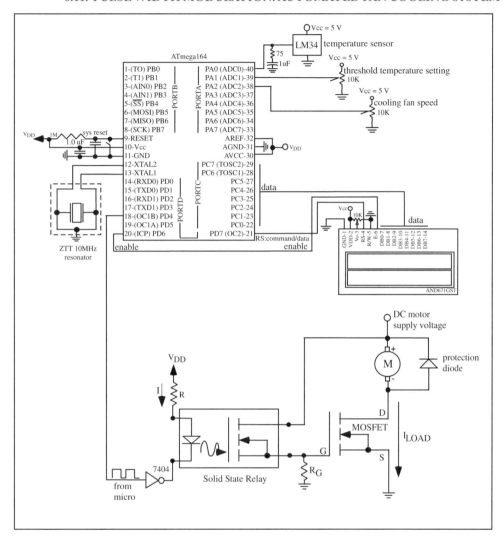

Figure 6.17: Automated fan cooling system.

```
//Last Updated: August 16, 2009
//****************************************************************************
//Description: This program reads the voltage from an LM34 temperature sensor
//then sends the corresponding temperature to an LCD. If the sensed
//temperature is greater than the temperature designated by a potentiometer,
//then a PWM signal is turned on to trigger a fan with duty cycle designated
//by another potentiometer.
```

```
//
//
//Ports:
//   PORTC[7:0]: data output to LCD
//   PORTD[7:6]: LCD control pins
//   PORTA[2:0]:
//   PORTA[0]: LM34 temperature sensor
//   PORTA[1]: threshold temperature
//   PORTA[2]: fan speed
//   PORTD[4]   : PWM channel B output
//
//****************************************************************************

//include files***************************************************************
#include<iom164pv.h>

//function prototypes*********************************************************
void initializePorts();
void initializeADC();
unsigned int readADC(unsigned char);
void LCD_init();
void putChar(unsigned char);
void putcommand(unsigned char);
void voltageToLCD(unsigned int);
void temperatureToLCD(unsigned int);
void PWM(unsigned int);
void delay_5ms();

int main(void)
{
unsigned int tempVoltage, tempThreshold;

initializePorts();
initializeADC();
LCD_init();

while(1)
   {
   tempVoltage = readADC(0);
```

```
  temperatureToLCD(tempVoltage);
  tempThreshold = readADC(1);
  if(tempVoltage > tempThreshold)
    {
    PWM(1);
    while(tempVoltage > tempThreshold)
      {
      tempVoltage = readADC(0);
      temperatureToLCD(tempVoltage);
      tempThreshold = readADC(1);
      }
    OCR1BL = 0x00;
    }
  }
return 0;
}


//***************************************************************************

void initializePorts()
{
DDRD = 0xFF;
DDRC = 0xFF;
DDRB = 0xFF;
}


//***************************************************************************

void initializeADC()
{
//select channel 0
ADMUX = 0;

//enable ADC and set module enable ADC and
//set module prescalar to 8
ADCSRA = 0xC3;

//Wait until conversion is ready
while(!(ADCSRA & 0x10));
```

```
//Clear conversion ready flag
ADCSRA |= 0x10;
}


//*************************************************************************


unsigned int readADC(unsigned char channel)
{
unsigned int binary_weighted_voltage, binary_weighted_voltage_low;
unsigned int binary_weighted_voltage_high; //weighted binary

ADMUX = channel;                         //Select channel
ADCSRA |= 0x43;                          //Start conversion


                                         //Set ADC module prescalar to 8
                                         //critical accurate ADC results
while (!(ADCSRA & 0x10));                 //Check if conversion is ready
ADCSRA |= 0x10;
 //Clear conv rdy flag - set the bit
binary_weighted_voltage_low = ADCL;
 //Read 8 low bits first (important)
   //Read 2 high bits, multiply by 256
binary_weighted_voltage_high = ((unsigned int)(ADCH << 8));
binary_weighted_voltage = binary_weighted_voltage_low +
                          binary_weighted_voltage_high;
return binary_weighted_voltage;          //ADCH:ADCL
}


//*************************************************************************
//LCD_Init: initialization for an LCD connected in the following manner:
//LCD: AND671GST 1x16 character display
//LCD configured as two 8 character lines in a 1x16 array
//LCD data bus (pin 14-pin7) ATMEL ATmega16: PORTC
//LCD RS (pin~(4) ATMEL ATmega16: PORTD[7]
//LCD E (pin~(6) ATMEL ATmega16: PORTD[6]
//*************************************************************************
```

```
void LCD_init(void)
{
delay_5ms();
delay_5ms();
delay_5ms();

                                    // output command string to
                                    //initialize LCD
putcommand(0x38);                   //function set 8-bit
delay_5ms();
putcommand(0x38);                   //function set 8-bit
delay_5ms();
putcommand(0x38);                   //function set 8-bit
putcommand(0x38);                   //one line, 5x7 char
putcommand(0x0E);                   //display on
putcommand(0x01);                   //display clear-1.64 ms
putcommand(0x06);                   //entry mode set
putcommand(0x00);                   //clear display, cursor at home
putcommand(0x00);                   //clear display, cursor at home
}


//****************************************************************************
//putchar:prints specified ASCII character to LCD
//****************************************************************************

void putChar(unsigned char~(c)
{
DDRC = 0xff; //set PORTC as output
DDRD = DDRD|0xC0; //make PORTD[7:6] output
PORTC = c;
PORTD = PORTD|0x80; //RS=1
PORTD = PORTD|0x40; //E=1
PORTD = PORTD&0xbf; //E=0
delay_5ms();
}


//****************************************************************************
//performs specified LCD related command
//****************************************************************************
```

```c
void putcommand(unsigned char~(d)
{
DDRC = 0xff; //set PORTC as output
DDRD = DDRD|0xC0; //make PORTD[7:6] output
PORTD = PORTD&0x7f; //RS=0
PORTC = d;
PORTD = PORTD|0x40; //E=1
PORTD = PORTD&0xbf; //E=0
delay_5ms();
}


//*******************************************************************************
//delays for 5 ms with a clock speed of 1 MHz
//*******************************************************************************

void delay_5ms(void)
{
unsigned int i;

for(i=0; i<2500; i++)
  {
  asm("nop");
  }
}


//*******************************************************************************
void voltageToLCD(unsigned int ADCValue)

{
float voltage;
unsigned int ones, tenths, hundredths;

voltage = (float)ADCValue*5.0/1024.0;

ones = (unsigned int)voltage;
tenths = (unsigned int)((voltage-(float)ones)*10);
hundredths = (unsigned int)(((voltage-(float)ones)*10-(float)tenths)*10);
```

```
putcommand(0x80);

putChar((unsigned char)(ones)+48);
putChar('.');
putChar((unsigned char)(tenths)+48);
putChar((unsigned char)(hundredths)+48);
putChar('V');
putcommand(0xC0);
}

//****************************************************************************

void temperatureToLCD(unsigned int ADCValue)
{
float voltage,temperature;
unsigned int tens, ones, tenths;

voltage = (float)ADCValue*5.0/1024.0;
temperature = voltage*100;

tens = (unsigned int)(temperature/10);
ones = (unsigned int)(temperature-(float)tens*10);
tenths = (unsigned int)(((temperature-(float)tens*10)-(float)ones)*10);

putcommand(0x80);
putChar((unsigned char)(tens)+48);
putChar((unsigned char)(ones)+48);
putChar('.');
putChar((unsigned char)(tenths)+48);
putChar('F');
}

//****************************************************************************

void PWM(unsigned int PWM_incr)
{
unsigned int fan_Speed_int;
float fan_Speed_float;
int PWM_duty_cycle;
```

```
fan_Speed_int = readADC(0x02); //fan Speed Setting

//unsigned int convert to max duty cycle setting:
//   0 VDC =  50% = 127,
//   5 VDC = 100% = 255

fan_Speed_float = ((float)(fan_Speed_int)/(float)(0x0400));

//convert volt to PWM constant 127-255
fan_Speed_int = (unsigned int)((fan_Speed_float * 127) + 128.0);

//Configure PWM clock
TCCR1A = 0xA1;
   //freq = resonator/510 = 4 MHz/510
                //freq = 19.607 kHz
TCCR1B = 0x02;                          //clock source
                                        //division of 8: 980 Hz


   //Initiate PWM duty cycle variables
PWM_duty_cycle = 0;
OCR1BH = 0x00;
OCR1BL = (unsigned char)(PWM_duty_cycle);//set PWM duty cycle Ch B to 0%
                                        //Ramp up to fan Speed in 1.6s
OCR1BL = (unsigned char)(PWM_duty_cycle);//set PWM duty cycle Ch B

while (PWM_duty_cycle < fan_Speed_int)
  {
  if(PWM_duty_cycle < fan_Speed_int)      //increment duty cycle
  PWM_duty_cycle=PWM_duty_cycle + PWM_incr;
  OCR1BL = (unsigned char)(PWM_duty_cycle);//set PWM duty cycle Ch B
  }
}


//************************************************************************
```

6.12   SUMMARY

In this chapter, we considered a microcontroller timer system, associated terminology for timer related topics, discussed typical functions of a timer subsystem, studied timer hardware operations, and considered some applications where the timer subsystem of a microcontroller can be used. We then took a detailed look at the timer subsystem aboard the ATmega164 and reviewed the features, operation, registers, and programming of the three timer channels. We concluded with examples employing a servo motor and an automated fan cooling system.

6.13 CHAPTER PROBLEMS

6.1. Given an 8 bit free running counter and the system clock rate of 24 MHz, find the time required for the counter to count from zero to its maximum value.

6.2. If we desire to generate periodic signals with periods ranging from 125 nanoseconds to 500 microseconds, what is the minimum frequency of the system clock?

6.3. Describe how you can compute the period of an incoming signal with varying duty cycles.

6.4. Describe how one can generate an aperiodic pulse with a pulse width of 2 minutes.

6.5. Program the output compare system of the ATmega164 to generate a 1 kHz signal with a 10 percent duty cycle.

6.6. Design a microcontroller system to control a sprinkler controller that performs the following tasks. We assume that your microcontroller runs with 10 MHz clock, and it has a 16 bit free running counter. The sprinkler controller system controls two different zones by turning sprinklers within each zone on and off. To turn on the sprinklers of a zone, the controller needs to receive a 152.589 Hz PWM signal from your microcontroller. To turn off the sprinklers of the same zone, the controller needs to receive the PWM signal with a different duty cycle.

6.7. Your microcontroller needs to provide the PWM signal with 10% duty cycle for 10 millisecond to turn on the sprinklers in zone one.

6.8. After 15 minutes, your microcontroller must send the PWM signal with 15% duty cycle for 10 millisecond to turn off the sprinklers in zone one.

6.9. After 15 minutes, your microcontroller must send the PWM signal with 20% duty cycle for 10 millisecond to turn on the sprinklers in zone two.

6.10. After 15 minutes, your microcontroller must send the PWM signal with 25% duty cycle for 10 millisecond to turn off the sprinklers in zone two.

6.11. Modify the servo motor example to include a potentiometer connected to PORTA[0]. The servo will deflect 0 degrees for 0 VDC applied to PORTA[0] and 180 degrees for 5 VDC.

6.12. For the automated cooling fan example, what would be the effect of changing the PWM frequency applied to the fan?

6.13. Modify the code of the automated cooling fan example to also display the set threshold temperature.

REFERENCES

Kenneth Short, *Embedded Microprocessor Systems Design: An Introduction Using the IN-TEL 80C188EB*, Prentice Hall, Upper Saddle River, 1998.

Frederick Driscoll, Robert Coughlin, and Robert Villanucci, *Data Acquisition and Process Control with the M68HC11 Microcontroller*, Second Edition, Prentice Hall, Upper Saddle River, 2000.

Todd Morton, *Embedded Microcontrollers*, Prentice Hall, Upper Saddle River, Prentice Hall, 2001.

Atmel 8-bit AVR Microcontroller with 16K Bytes In-System Programmable Flash, ATmega164, AT-mega164L, data sheet: 2466L-AVR-06/05, Atmel Corporation, 2325 Orchard Parkway, San Jose, CA 95131.

S. Barrett and D. Pack (2006) Microcontrollers Fundamentals for Engineers and Scientists. Morgan and Claypool Publishers.

CHAPTER 7

Atmel AVR Operating Parameters and Interfacing

Objectives: After reading this chapter, the reader should be able to

- Describe the voltage and current parameters for the Atmel AVR HC CMOS type microcontroller.

- Specify a battery system to power an Atmel AVR based system.

- Apply the voltage and current parameters toward properly interfacing input and output devices to the Atmel AVR microcontroller.

- Interface a wide variety of input and output devices to the Atmel AVR microcontroller.

- Describe the special concerns that must be followed when the Atmel AVR microcontroller is used to interface to a high power DC or AC device.

- Discuss the requirement for an optical based interface.

- Describe how to control the speed and direction of a DC motor.

- Describe how to control several types of AC loads.

The textbook for Morgan & Claypool Publishers (M&C) titled, "Microcontrollers Fundamentals for Engineers and Scientists," contains a chapter entitled "Operating Parameters and Interfacing." With M&C permission, we repeated portions of the chapter here for your convenience. However, we have customized the information provided to the Atmel AVR line of microcontrollers and have also expanded the coverage of the chapter to include interface techniques for a number of additional input and output devices.

In this chapter, we introduce you to the extremely important concepts of the operating envelope for a microcontroller. We begin by reviewing the voltage and current electrical parameters for the HC CMOS based Atmel AVR line of microcontrollers. We then show how to apply this information to properly interface input and output devices to the ATmega164 microcontroller. We then discuss the special considerations for controlling a high power DC or AC load such as a motor and introduce the concept of an optical interface. Throughout the chapter, we provide a number of detailed examples.

The importance of this chapter can not be emphasized enough. Any time an input or an output device is connected to a microcontroller, the interface between the device and the microcontroller must be carefully analyzed and designed. This will ensure the microcontroller will continue to operate within specified parameters. Should the microcontroller be operated outside its operational envelope, erratic, unpredictable, and unreliable system may result.

7.1   OPERATING PARAMETERS

Any time a device is connected to a microcontroller, careful interface analysis must be performed. Most microcontrollers are members of the "HC," or high-speed CMOS, family of chips. As long as all components in a system are also of the "HC" family, as is the case for the Atmel AVR line of microcontrollers, electrical interface issues are minimal. If the microcontroller is connected to some component not in the "HC" family, electrical interface analysis must be completed. Manufacturers readily provide the electrical characteristic data necessary to complete this analysis in their support documentation.

To perform the interface analysis, there are eight different electrical specifications required for electrical interface analysis. The electrical parameters are:

- V_{OH}: the lowest guaranteed output voltage for a logic high,

- V_{OL}: the highest guaranteed output voltage for a logic low,

- I_{OH}: the output current for a V_{OH} logic high,

- I_{OL}: the output current for a V_{OL} logic low,

- V_{IH}: the lowest input voltage guaranteed to be recognized as a logic high,

- V_{IL}: the highest input voltage guaranteed to be recognized as a logic low,

- I_{IH}: the input current for a V_{IH} logic high, and

- I_{IL}: the input current for a V_{IL} logic low.

These electrical characteristics are required for both the microcontroller and the external components. Typical values for a microcontroller in the HC CMOS family assuming $V_{DD} = 5.0$ volts and $V_{SS} = 0$ volts are provided below. The minus sign on several of the currents indicates a current flow out of the device. A positive current indicates current flow into the device.

- $V_{OH} = 4.2$ volts,

- $V_{OL} = 0.4$ volts,

- $I_{OH} = -0.8$ milliamps,

- $I_{OL} = 1.6$ milliamps,

- $V_{IH} = 3.5$ volts,

- $V_{IL} = 1.0$ volt,

- $I_{IH} = 10$ microamps, and

- $I_{IL} = -10$ microamps.

It is important to realize that these are static values taken under very specific operating conditions. If external circuitry is connected such that the microcontroller acts as a current source (current leaving the microcontroller) or current sink (current entering the microcontroller), the voltage parameters listed above will also be affected.

In the current source case, an output voltage V_{OH} is provided at the output pin of the microcontroller when the load connected to this pin draws a current of I_{OH}. If a load draws more current from the output pin than the I_{OH} specification, the value of V_{OH} is reduced. If the load current becomes too high, the value of V_{OH} falls below the value of V_{IH} for the subsequent logic circuit stage and not be recognized as an acceptable logic high signal. When this situation occurs, erratic and unpredictable circuit behavior results.

In the sink case, an output voltage V_{OL} is provided at the output pin of the microcontroller when the load connected to this pin delivers a current of I_{OL} to this logic pin. If a load delivers more current to the output pin of the microcontroller than the I_{OL} specification, the value of V_{OL} increases. If the load current becomes too high, the value of V_{OL} rises above the value of V_{IL} for the subsequent logic circuit stage and not be recognized as an acceptable logic low signal. As before, when this situation occurs, erratic and unpredictable circuit behavior results.

For convenience, this information is illustrated in Figure 7.1. In (a), we provided an illustration of the direction of current flow from the HC device and also a comparison of voltage levels. As a reminder, current flowing out of a device is considered a negative current (source case) while current flowing into the device is considered positive current (sink case). The magnitude of the voltage and current for HC CMOS devices are shown in (b). As more current is sinked or sourced from a microcontroller pin, the voltage will be pulled up or pulled down, respectively, as shown in (c). If input and output devices are improperly interfaced to the microcontroller, these loading conditions may become excessive and voltages will not be properly interpreted as the correct logic levels.

You must also ensure that total current limits for an entire microcontroller port and overall bulk port specifications are complied with. For planning purposes, the sum of current sourced or sinked from a port should not exceed 100 mA. Furthermore, the sum of currents for all ports should not exceed 200 mA. As before, if these guidelines are not followed, erratic microcontroller behavior may result.

The procedures presented in the following sections, when followed carefully, will ensure the microcontroller will operate within its designed envelope. The remainder of the chapter is divided into input device interface analysis followed by output device interface analysis. Since many embedded systems operate from a DC battery source, we begin by examining several basic battery supply circuits.

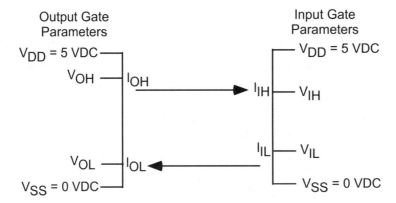

a) Voltage and current electrical parameters

Output Parameters	Input Parameters
$V_{OH} = 4.2$ V	$V_{IH} = 3.5$ V
$V_{OL} = 0.4$ V	$V_{IL} = 1.0$ V
$I_{OH} = -0.8$ mA	$I_{IH} = 10$ μA
$I_{OL} = 1.6$ mA	$I_{IL} = -10$ μA

b) HC CMOS voltage and current parameters

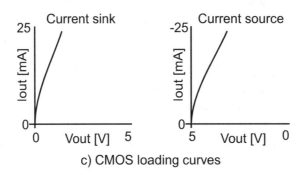

c) CMOS loading curves

Figure 7.1: Electrical voltage and current parameters.

7.2   BATTERY OPERATION

Many embedded systems are remote, portable systems operating from a battery supply. To properly design a battery source for an embedded system, the operating characteristics of the embedded system must be matched to the characteristics of the battery supply.

7.2.1   EMBEDDED SYSTEM VOLTAGE AND CURRENT DRAIN SPECIFICA-TIONS

An embedded system has a required supply voltage and an overall current requirement. For the purposes of illustration, we will assume our microcontroller based embedded system operates from 5 VDC. The overall current requirements of the system is determined by the worst case current requirements when all embedded system components are operational.

7.2.2   BATTERY CHARACTERISTICS

To properly match a battery to an embedded system, the battery voltage and capacity must be specified. Battery capacity is typically specified as a mAH rating. For example, a typical 9 VDC non-rechargeable alkaline battery has a capacity of 550 mAH. If the embedded system has a maximum operating current of 50 mA, it will operate for approximately eleven hours before battery replacement is required.

A battery is typically used with a voltage regulator to maintain the voltage at a prescribed level. Figure 7.2 provides sample circuits to provide a +5 VDC and a ±5 VDC portable battery source. Additional information on battery capacity and characteristics may be found in Barrett and Pack (S. Barrett, 2004).

7.3   INPUT DEVICES

In this section, we discuss how to properly interface input devices to a microcontroller. We will start with the most basic input component, a simple on/off switch.

7.3.1   SWITCHES

Switches come in a variety of types. As a system designer it is up to you to choose the appropriate switch for a specific application. Switch varieties commonly used in microcontroller applications are illustrated in Figure 7.3(a). Here is a brief summary of the different types:

- **Slide switch:** A slide switch has two different positions: on and off. The switch is manually moved to one position or the other. For microcontroller applications, slide switches are available that fit in the profile of a common integrated circuit size dual inline package (DIP). A bank of four or eight DIP switches in a single package is commonly available.

- **Momentary contact pushbutton switch:** A momentary contact pushbutton switch comes in two varieties: normally closed (NC) and normally open (NO). A normally open switch,

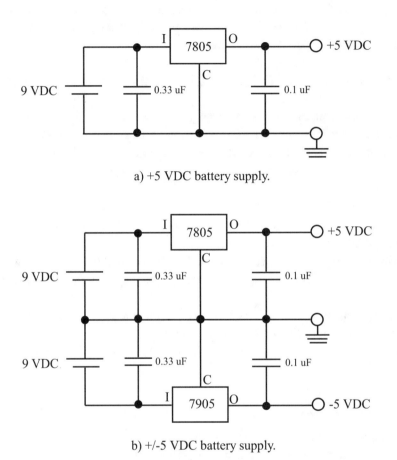

a) +5 VDC battery supply.

b) +/-5 VDC battery supply.

Figure 7.2: Battery supply circuits employing a 9 VDC battery with a 5 VDC regulators.

as its name implies, does not normally provide an electrical connection between its contacts. When the pushbutton portion of the switch is depressed, the connection between the two switch contacts is made. The connection is held as long as the switch is depressed. When the switch is released, the connection is opened. The converse is true for a normally closed switch. For microcontroller applications pushbutton switches are available in a small tact type switch configuration.

- **Push on/push off switches:** These types of switches are also available in a normally open or normally closed configuration. For the normally open configuration, the switch is depressed to make connection between the two switch contacts. The pushbutton must be depressed again to release the connection.

- **Hexadecimal rotary switches:** Small profile rotary switches are available for microcontroller applications. These switches commonly have sixteen rotary switch positions. As the switch is rotated to each position, a unique four bit binary code is provided at the switch contacts.

A common switch interface is shown in Figure 7.3(b). This interface allows a logic one or zero to be properly introduced to a microcontroller input port pin. The basic interface consists of the switch in series with a current limiting resistor. The node between the switch and the resistor is provided to the microcontroller input pin. In the configuration shown, the resistor pulls the microcontroller input up to the supply voltage V_{DD}. When the switch is closed, the node is grounded and a logic zero is provided to the microcontroller input pin. To reverse the logic of the switch configuration the position of the resistor and the switch is simply reversed.

7.3.2   PULLUP RESISTORS IN SWITCH INTERFACE CIRCUITRY

Many microcontrollers are equipped with pullup resistors at the input pins. The pullup resistors are asserted with the appropriate register setting. The pullup resistor replaces the external resistor in the switch configuration as shown in Figure 7.3b) right.

7.3.3   SWITCH DEBOUNCING

Mechanical switches do not make a clean transition from one position (on) to another (off). When a switch is moved from one position to another, it makes and breaks contact multiple times. This activity may go on for tens of milliseconds. A microcontroller is relatively fast as compared to the action of the switch. Therefore, the microcontroller is able to recognize each switch bounce as a separate and erroneous transition.

To correct the switch bounce phenomena, additional external hardware components may be used or software techniques may be employed. A hardware debounce circuit is illustrated in Figure 7.3(c). The node between the switch and the limiting resistor of the basic switch circuit is fed to a low pass filter (LPF) formed by the 470K ohm resistor and the capacitor. The LPF prevents abrupt changes (bounces) in the input signal from the microcontroller. The LPF is followed by a 74HC14 Schmitt Trigger, which is simply an inverter equipped with hysteresis. This further limits the switch bouncing.

Switches may also be debounced using software techniques. This is accomplished by inserting a 30 to 50 ms lockout delay in the function responding to port pin changes. The delay prevents the microcontroller from responding to the multiple switch transitions related to bouncing.

You must carefully analyze a given design to determine if hardware or software switch debouncing techniques will be used. It is important to remember that all switches exhibit bounce phenomena and, therefore, must be debounced.

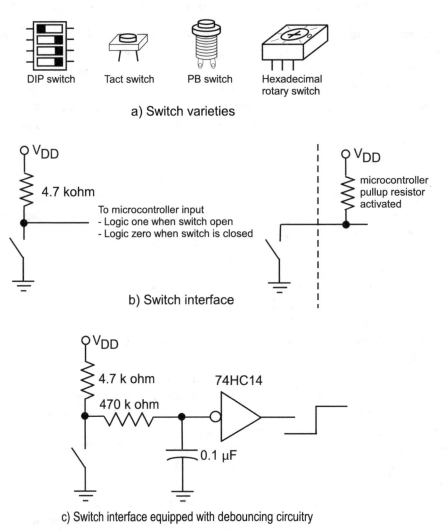

Figure 7.3: Switch interface.

7.3.4   KEYPADS

A keypad is simply an extension of the simple switch configuration. A typical keypad configuration and interface are shown in Figure 7.4. As you can see, the keypad is simply multiple switches in the same package. A hexadecimal keypad is provided in the figure. A single row of keypad switches are asserted by the microcontroller and then the host keypad port is immediately read. If a switch has been depressed, the keypad pin corresponding to the column the switch is in will also be asserted. The combination of a row and a column assertion can be decoded to determine which key has been pressed as illustrated in the table. Keypad rows are continually asserted one after the other in sequence. Since the keypad is a collection of switches, debounce techniques must also be employed.

The keypad may be used to introduce user requests to a microcontroller. A standard keypad with alphanumeric characters may be used to provide alphanumeric values to the microcontroller such as providing your personal identification number (PIN) for a financial transaction. However, some keypads are equipped with removable switch covers such that any activity can be associated with a key press.

In Figure 7.5, we have connected the ATmega164 to a hexadecimal keypad via PORTC. PORTC[3:0] is configured as output to selectively assert each row. PORTC[7:4] is configured as input. Each row is sequentially asserted low. Each column is then read via PORTC[7:4] to see if any switch in that row has been depressed. If no switches have been depressed in the row, an "F" will be read from PORTC[7:4]. If a switch has been depressed, some other value than "F" will be read. The read value is then passed into a switch statement to determine the ASCII equivalent of the depressed switch. The function is not exited until the switch is released. This prevents a switch "double hit."

```
//*************************************************************************

unsigned char get_keypad_value(void)
{
unsigned char  PORTC_value, PORTC_value_masked;
unsigned char  ascii_value;

DDRC = 0x0F;
        //set PORTC[7:4] to input,
                                        //PORTC[3:0] to output

                                        //switch depressed in row 0?
PORTC = 0xFE;                           //assert row 0 via PORTC[0]
PORTC_value = PINC;                     //read PORTC
PORTC_value_masked = (PORTC_value & 0xf0);  //mask PORTC[3:0]

                                        //switch depressed in row 1?
```

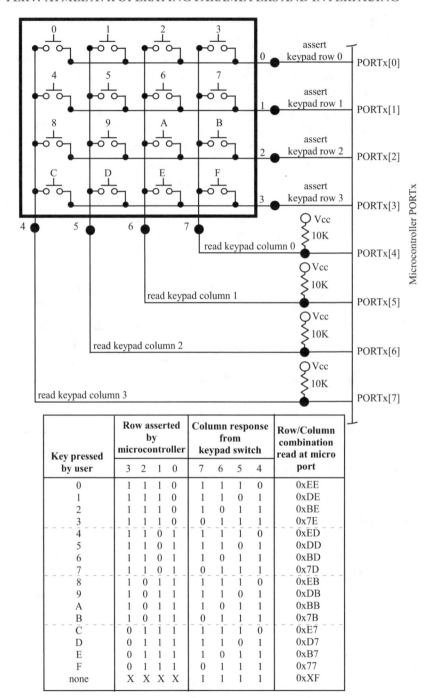

Key pressed by user	Row asserted by microcontroller				Column response from keypad switch				Row/Column combination read at micro port
	3	2	1	0	7	6	5	4	
0	1	1	1	0	1	1	1	0	0xEE
1	1	1	1	0	1	1	0	1	0xDE
2	1	1	1	0	1	0	1	1	0xBE
3	1	1	1	0	0	1	1	1	0x7E
4	1	1	0	1	1	1	1	0	0xED
5	1	1	0	1	1	1	0	1	0xDD
6	1	1	0	1	1	0	1	1	0xBD
7	1	1	0	1	0	1	1	1	0x7D
8	1	0	1	1	1	1	1	0	0xEB
9	1	0	1	1	1	1	0	1	0xDB
A	1	0	1	1	1	0	1	1	0xBB
B	1	0	1	1	0	1	1	1	0x7B
C	0	1	1	1	1	1	1	0	0xE7
D	0	1	1	1	1	1	0	1	0xD7
E	0	1	1	1	1	0	1	1	0xB7
F	0	1	1	1	0	1	1	1	0x77
none	X	X	X	X	1	1	1	1	0xXF

Figure 7.4: Keypad interface.

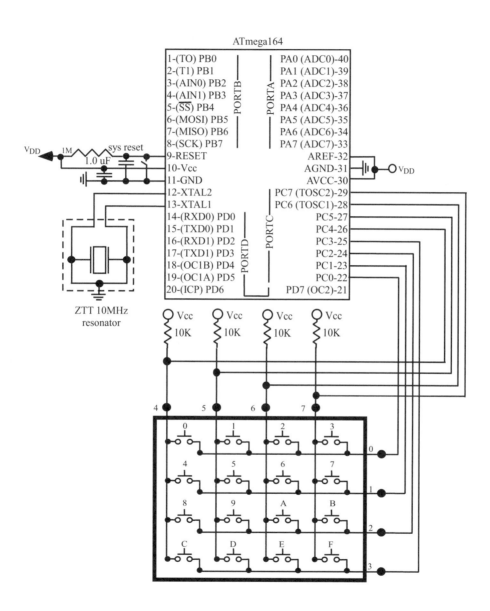

Figure 7.5: Hexadecimal keypad interface to microcontroller.

```
if(PORTC_value_masked == 0xf0)
      //no switches depressed in row 0
   {
   PORTC = 0xFD;                            //assert Row 1 via PORTC[1]
   PORTC_value = PINC;                      //read PORTC
   PORTC_value_masked = (PORTC_value & 0xf0);//mask PORTC[3:0]
   }


                                           //switch depressed in row 2?
if(PORTC_value_masked == 0xf0)
      //no switches depressed in row 0
   {
   PORTC = 0xFB;                            //assert Row 2 via PORTC[2]
   PORTC_value = PINC;                      //read PORTC
   PORTC_value_masked = (PORTC_value & 0xf0);//mask PORTC[3:0]
   }


                                           //switch depressed in row 3?
if(PORTC_value_masked == 0xf0)
      //no switches depressed in row 0
   {
   PORTC = 0xF7;                            //assert Row 3 via PORTC[3]
   PORTC_value = PINC;                      //read PORTC
   PORTC_value_masked = (PORTC_value & 0xf0);//mask PORTC[3:0]
   }

if(PORTC_value_masked != 0xf0)
  {
  switch(PORTC_value_masked)
    {
    case 0xEE: ascii_value = '0';
               break;

    case 0xDE: ascii_value = '1';
               break;

    case 0xBE: ascii_value = '2';
               break;
```

```
case 0x7E: ascii_value = '3';
           break;

case 0xED: ascii_value = '4';
           break;

case 0xDD: ascii_value = '5';
           break;

case 0xBD: ascii_value = '6';
           break;

case 0x7D: ascii_value = '7';
           break;

case 0xEB: ascii_value = '8';
           break;

case 0xDB: ascii_value = '9';
           break;

case 0xBB: ascii_value = 'a';
           break;

case 0x&B: ascii_value = 'b';
           break;

case 0xE7: ascii_value = 'c';
           break;

case 0xD7: ascii_value = 'd';
           break;

case 0xB7: ascii_value = 'e';
           break;

case 0x77: ascii_value = 'f';
           break;
```

```
      default:;
      }

while(PORTC_value_masked != 0xf0);
        //wait for key to be released

return ascii_value;
}


//***********************************************************************
```

7.3.5   SENSORS

A microcontroller is typically used in control applications where data is collected, the data is assimilated and processed by the host algorithm, and a control decision and accompanying signals are provided by the microcontroller. Input data for the microcontroller is collected by a complement of input sensors. These sensors may be digital or analog in nature.

7.3.5.1 Digital Sensors

Digital sensors provide a series of digital logic pulses with sensor data encoded. The sensor data may be encoded in any of the parameters associated with the digital pulse train such as duty cycle, frequency, period, or pulse rate. The input portion of the timing system may be configured to measure these parameters.

An example of a digital sensor is the optical encoder. An optical encoder consists of a small plastic transparent disk with opaque lines etched into the disk surface. A stationary optical emitter and detector pair is placed on either side of the disk. As the disk rotates, the opaque lines break the continuity between the optical source and detector. The signal from the optical detector is monitored to determine disk rotation as shown in Figure 7.6.

Optical encoders are available in a variety of types depending on the information desired. There are two major types of optical encoders: incremental encoders and absolute encoders. An absolute encoder is used when it is required to retain position information when power is lost. For example, if you were using an optical encoder in a security gate control system, an absolute encoder would be used to monitor the gate position. An incremental encoder is used in applications where a velocity or a velocity and direction information is required.

The incremental encoder types may be further subdivided into tachometers and quadrature encoders. An incremental tachometer encoder consists of a single track of etched opaque lines as shown in Figure 7.6(a). It is used when the velocity of a rotating device is required. To calculate velocity the number of detector pulses are counted in a fixed amount of time. Since the number of pulses per encoder revolution is known, velocity may be calculated.

The quadrature encoder contains two tracks shifted in relationship to one another by 90 degrees. This allows the calculation of both velocity and direction. To determine direction, one would

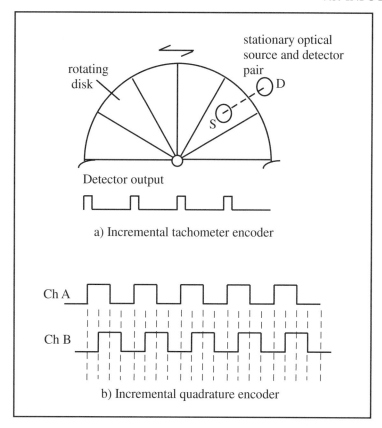

rotating disk

stationary optical source and detector pair

D

S

Detector output

a) Incremental tachometer encoder

Ch A

Ch B

b) Incremental quadrature encoder

Figure 7.6: Optical encoder.

monitor the phase relationship between Channel A and Channel B as shown in Figure 7.6(b). The absolute encoder is equipped with multiple data tracks to determine the precise location of the encoder disk (Sick).

Example: Optical encoder for motor speed measurement and control. In Chapter 8, we provide a detailed example of actively regulating the speed of a motor under various loads. An optical tachometer is used to monitor instantaneous motor speed. Based on the speed measurement, the motor is incrementally sped up or slowed down by varying the duty cycle of the motor PWM signal.

7.3.5.2 Analog Sensors

Analog sensors provide a DC voltage that is proportional to the physical parameter being measured. As discussed in the analog to digital conversion chapter, the analog signal may be first preprocessed by external analog hardware such that it falls within the voltage references of the conversion subsystem. The analog voltage is then converted to a corresponding binary representation.

An example of an analog sensor is the flex sensor shown in Figure 7.7(a). The flex sensor provides a change in resistance for a change in sensor flexure. At 0 degrees flex, the sensor provides 10K ohms of resistance. For 90 degrees flex, the sensor provides 30-40K ohms of resistance. Since the microcontroller can not measure resistance directly, the change in flex sensor resistance must be converted to a change in a DC voltage. This is accomplished using the voltage divider network shown in Figure 7.7(c). For increased flex, the DC voltage will increase. The voltage can be measured using the ATmega164's analog to digital converter subsystem. The flex sensor may be used in applications such as virtual reality data gloves, robotic sensors, biometric sensors, and in science and engineering experiments (Images).

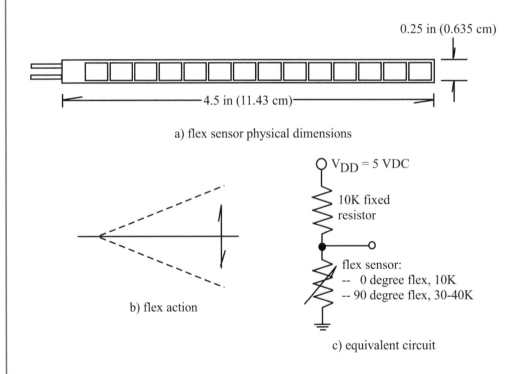

a) flex sensor physical dimensions

b) flex action

c) equivalent circuit

Figure 7.7: Flex sensor.

7.3.6   LM34 TEMPERATURE SENSOR EXAMPLE

Temperature may be sensed using an LM34 (Fahrenheit) or LM35 (Centigrade) temperature transducer. The LM34 provides an output voltage that is linearly related to temperature. For example, the LM34D operates from 32 degrees F to 212 degrees F providing +10mV/degree Fahrenheit resolution with a typical accuracy of ± 0.5 degrees Fahrenheit (National). This sensor was used in the automated cooling fan example of Chapter 6. The output from the sensor is typically connected to the ADC input of the microcontroller.

7.4   OUTPUT DEVICES

As previously mentioned, an external device should not be connected to a microcontroller without first performing careful interface analysis to ensure the voltage, current, and timing requirements of the microcontroller and the external device. In this section, we describe interface considerations for a wide variety of external devices. We begin with the interface for a single light emitting diode.

7.4.1   LIGHT EMITTING DIODES (LEDS)

An LED is typically used as a logic indicator to inform the presence of a logic one or a logic zero at a specific pin of a microcontroller. An LED has two leads: the anode or positive lead and the cathode or negative lead. To properly bias an LED, the anode lead must be biased at a level approximately 1.7 to 2.2 volts higher than the cathode lead. This specification is known as the forward voltage (V_f) of the LED. The LED current must also be limited to a safe level known as the forward current (I_f). The diode voltage and current specifications are usually provided by the manufacturer.

An example of an LED biasing circuit is provided in Figure 7.8. A logic one is provided by the microcontroller to the input of the inverter. The inverter provides a logic zero at its output which provides a virtual ground at the cathode of the LED. Therefore, the proper voltage biasing for the LED is provided. The resistor (R) limits the current through the LED. A proper resistor value can be calculated using $R = (V_{DD} - V_{DIODE})/I_{DIODE}$. It is important to note that a 7404 inverter must be used due to its capability to safely sink 16 mA of current. Alternately, an NPN transistor such as a 2N2222 (PN2222 or MPQ2222) may be used in place of the inverter as shown in the figure. In Chapter 1, we used large (10 mm) red LEDs in the KNH instrumentation project. These LEDs have V_f of 6 to 12 VDC and I_f of 20 mA at 1.85 VDC. This requires the interface circuit shown in Figure 7.8c) right.

7.4.2   SEVEN SEGMENT LED DISPLAYS

To display numeric data, seven segment LED displays are available as shown in Figure 7.9(a). Different numerals can be displayed by asserting the proper LED segments. For example, to display the number five, segments a, c, d, f, and g would be illuminated. Seven segment displays are available in common cathode (CC) and common anode (CA) configurations. As the CC designation implies, all seven individual LED cathodes on the display are tied together.

The microcontroller is not capable of driving the LED segments directly. As shown in Figure 7.9(a), an interface circuit is required. We use a 74LS244 octal buffer/driver circuit to boost the current available for the LED. The LS244 is capable of providing 15 mA per segment (I_{OH}) at 2.0 VDC (V_{OH}). A limiting resistor is required for each segment to limit the current to a safe

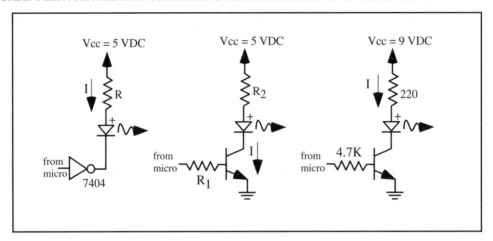

Figure 7.8: LED display devices.

value for the LED. Conveniently, resistors are available in DIP packages of eight for this type of application.

Seven segment displays are available in multi-character panels. In this case, separate micro-controller ports are not used to provide data to each seven segment character. Instead, a single port is used to provide character data. A portion of another port is used to sequence through each of the characters as shown in Figure 7.9(b). An NPN (for a CC display) transistor is connected to the common cathode connection of each individual character. As the base contact of each transistor is sequentially asserted, the specific character is illuminated. If the microcontroller sequences through the display characters at a rate greater than 30 Hz, the display will have steady illumination.

7.4.3   CODE EXAMPLE
Provided below is a function used to illuminate the correct segments on a multi-numeral seven display. The numeral is passed in as an argument to the function along with the numerals position on the display and also an argument specifying whether or not the decimal point (dp) should be displayed at that position. The information to illuminate specific segments are provided in Figure 7.9c).

```
//*********************************************************************
void LED_character_display(unsigned int numeral, unsigned int position,
                           unsigned int decimal_point)
{
unsigned char output_value;
                                //illuminate numerical segments
switch(numeral)
  {
```

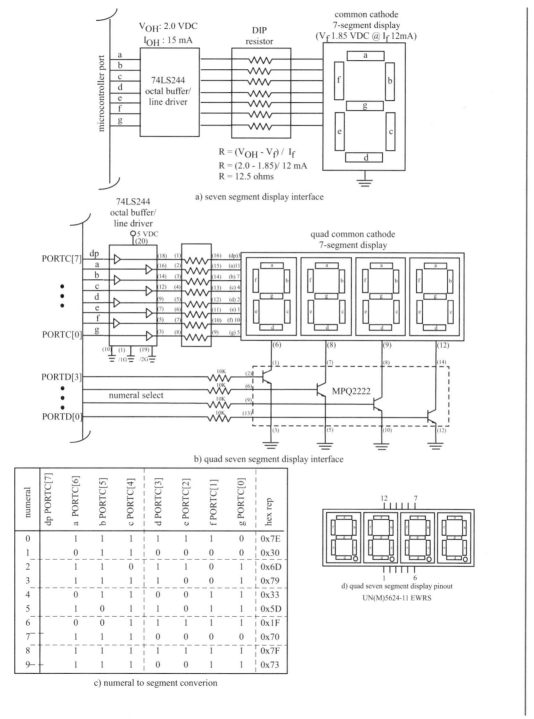

Figure 7.9: Seven segment LED display devices.

```
   case 0: output_value = 0x7E;
     break;

   case 1: output_value = 0x30;
     break;

   case 2: output_value = 0x6D;
     break;

   case 3: output_value = 0x79;
     break;

   case 4: output_value = 0x33;
     break;

   case 5: output_value = 0x5D;
     break;

   case 6: output_value = 0x1F;
     break;

   case 7: output_value = 0x70;
     break;

   case 8: output_value = 0x7F;
     break;

   case 9: output_value = 0x73;
     break;

   default:;
   }

if(decimal_point != 0)
   PORTC = output_value | 0x80;          //illuminate decimal point

switch(position)                         //assert position
   {
   case 0: PORTD = 0x01;                  //least significant bit
```

```
            break;

   case 1: PORTD = 0x02;                 //least significant bit + 1
           break;

   case 2: PORTD = 0x04;                 //least significant bit + 2
           break;

   case 3: PORTD = 0x08;                 //most significant bit
           break;

   default:;
   }
}
//********************************************************************
```

7.4.4   TRI-STATE LED INDICATOR

The tri-state LED indicator introduced in Chapter 2 is shown in Figure 7.10. It is used to provide the status of an entire microcontroller port. The indicator bank consists of eight green and eight red LEDs. When an individual port pin is logic high, the green LED is illuminated. When logic low, the red LED is illuminated. If the port pin is at a tri-state high impedance state, no LED is illuminated.

The NPN/PNP transistor pair at the bottom of the figure provides a 2.5 VDC voltage reference for the LEDs. When a specific port pin is logic high (5.0 VDC), the green LED will be forward biased since its anode will be at a higher potential than its cathode. The 47 ohm resistor limits current to a safe value for the LED. Conversely, when a specific port pin is at a logic low (0 VDC), the red LED will be forward biased and illuminate. For clarity, the red and green LEDs are shown as being separate devices. LEDs are available that have both LEDs in the same device.

7.4.5   DOT MATRIX DISPLAY

The dot matrix display consists of a large number of LEDs configured in a single package. A typical 5×7 LED arrangement is a matrix of five columns of LEDs with seven LEDs per row as shown in Figure 7.11. Display data for a single matrix column [R6-R0] is provided by the microcontroller. That specific row is then asserted by the microcontroller using the column select lines [C2-C0]. The entire display is sequentially built up a column at a time. If the microcontroller sequences through each column fast enough (greater than 30 Hz), the matrix display appears to be stationary to a human viewer.

In Figure 7.11a), we have provided the basic configuration for the dot matrix display for a single display device. However, this basic idea can be expanded in both dimensions to provide a multi-

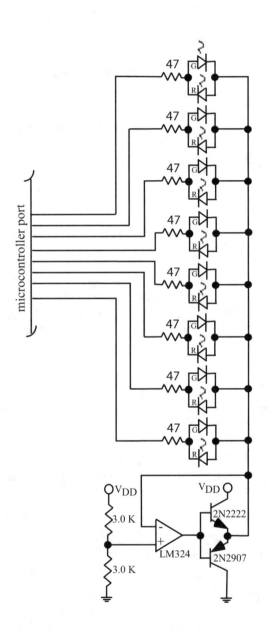

Figure 7.10: Tri-state LED display.

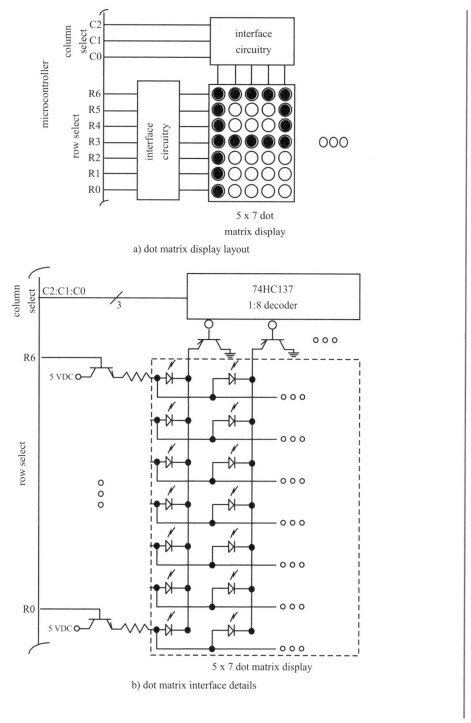

a) dot matrix display layout

b) dot matrix interface details

Figure 7.11: Dot matrix display.

character, multi-line display. A larger display does not require a significant number of microcontroller pins for the interface. The dot matrix display may be used to display alphanumeric data as well as graphics data. In Figure 7.11b), we have provided additional detail of the interface circuit.

7.4.6   LIQUID CRYSTAL CHARACTER DISPLAY (LCD)

An LCD is an output device to display text information as shown in Figure 7.12. LCDs come in a wide variety of configurations, including multi-character and multi-line format. A $16 times 2$ LCD format is common. That is, it has the capability of displaying two lines of 16 characters each. The characters are sent to the LCD via American Standard Code for Information Interchange (ASCII) format a single character at a time. For a parallel configured LCD, an eight bit data path and two lines are required between the microcontroller and the LCD. A small microcontroller mounted to the back panel of the LCD translates the ASCII data characters and control signals to properly display the characters. LCDs are configured for either parallel or serial data transmission format. In the example provided we use a parallel configured display. In Figure 7.13, we have included the LCD in the Testbench hardware configuration.

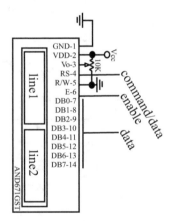

Figure 7.12: LCD display.

Some sample C code is provided below to send data and control signals to an LCD. In this specific example, an AND671GST 1 × 16 character LCD was connected to the Atmel ATmega164 microcontroller. One 8-bit port and two extra control lines are required to connect the microcontroller to the LCD. Note: The initialization sequence for the LCD is specified within the manufacturer's technical data.

```
//***************************************************************************
//LCD_Init: initialization for an LCD connected in the following manner:
//LCD: AND671GST 1x16 character display
//LCD configured as two 8 character lines in a 1x16 array
```

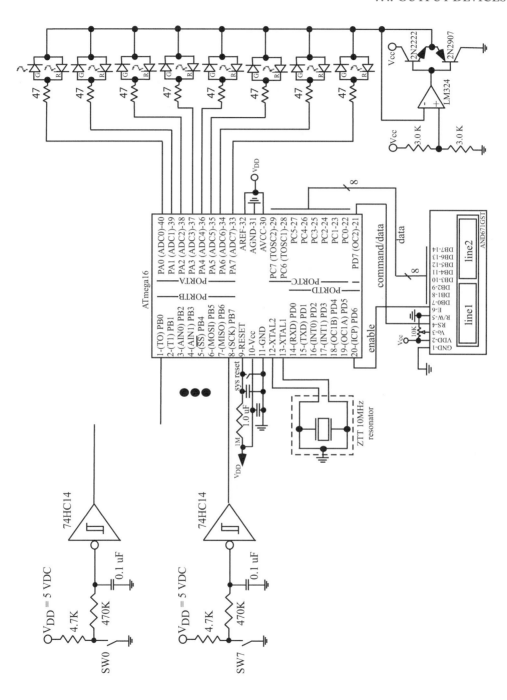

Figure 7.13: Hardware Testbench equipped with an LCD.

```
//LCD data bus (pin 14-pin7)   ATMEL ATmega164: PORTC
//LCD RS (pin~(4) ATMEL ATmega164: PORTD[7]


//LCD E  (pin~(6) ATMEL ATmega164: PORTD[6]
//****************************************************************

void LCD_Init(void)
{
delay_5ms();
delay_5ms();
delay_5ms();
                          // output command string to initialize LCD
putcommand(0x38);         //function set 8-bit
delay_5ms();
putcommand(0x38);         //function set 8-bit
putcommand(0x38);         //function set 8-bit
putcommand(0x38);         //one line, 5x7 char

putcommand(0x0C);         //display on

putcommand(0x01);         //display clear-1.64 ms
putcommand(0x06);         //entry mode set
putcommand(0x00);         //clear display, cursor at home
putcommand(0x00);         //clear display, cursor at home
}


//****************************************************************
//putchar:prints specified ASCII character to LCD
//****************************************************************

void putchar(unsigned char~(c)
{
DDRC = 0xff;              //set PORTC as output
DDRD = DDRD|0xC0;         //make PORTD[7:6] output
PORTC = c;
PORTD = PORTD|0x80;       //RS=1
PORTD = PORTD|0x40;       //E=1
PORTD = PORTD&0xbf;       //E=0
delay_5ms();
```

```
}


//*************************************************************************
//performs specified LCD related command
//*************************************************************************


void putcommand(unsigned char~(d)
{
DDRC = 0xff;                     //set PORTC as output
DDRD = DDRD|0xC0;                //make PORTD[7:6] output
PORTD = PORTD&0x7f;              //RS=0
PORTC = d;
PORTD = PORTD|0x40;              //E=1
PORTD = PORTD&0xbf;              //E=0
delay_5ms();
}


//*************************************************************************
```

7.4.7   GRAPHIC LIQUID CRYSTAL DISPLAY (GLCD)

A graphic LCD may be used to generate custom displays. A display is assembled as a collection of picture elements (pixels). The pixels may be pre-arranged into specific graphic representations such as geometric shapes, fonts, or any custom image feature. In this section, we illustrate how to interface a graphic LCD to the ATmega164 and also how to program a specific pixel location on the GLCD. We employ the Hantronix HDM64GS12L-4 128×64 pixel graphics LCD display module as a representative sample of this type of display. Figures and sample code in this section were adapted from Hantronix literature and also sample GLCD code (Hantronix).

Interfacing a GLCD is similar to interfacing a character LCD. A number of control signals are required to control the action of the GLCD. We will use PORTD[5:0] of the ATmega164 to generate the control signals required by the GLCD. We employ PORTC of the ATmega164 to send data or instructions to the GLCD. The interface is illustrated in Figure 7.14. The definitions for each of the control signals are provided in Figure 7.15 along with the necessary timing signals and pin definitions for the GLCD.

The GLCD consists of two separate display areas designated left and right. Each display area is divided into eight horizontal pages designated page 1 through page 8. Each page consists of 64×8 bits of RAM memory. The left page is organized into 64 columns, (1 through 64) in the y direction and 64 segments (1 through 64) in the x direction. The right page is organized into 64 columns, (1 through 64) in the y direction and 64 segments (65 through 128) in the x direction. This organization is shown in Figure 7.15.

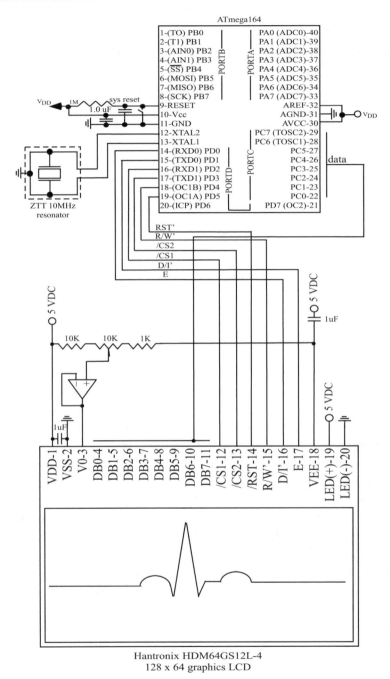

Figure 7.14: ATmega164 to graphic LCD interface.

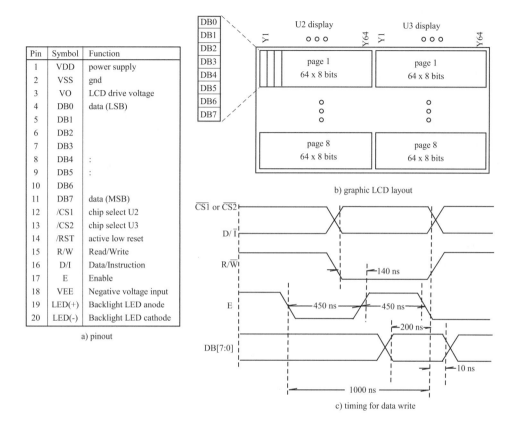

Pin	Symbol	Function
1	VDD	power supply
2	VSS	gnd
3	VO	LCD drive voltage
4	DB0	data (LSB)
5	DB1	
6	DB2	
7	DB3	
8	DB4	:
9	DB5	:
10	DB6	
11	DB7	data (MSB)
12	/CS1	chip select U2
13	/CS2	chip select U3
14	/RST	active low reset
15	R/W	Read/Write
16	D/I	Data/Instruction
17	E	Enable
18	VEE	Negative voltage input
19	LED(+)	Backlight LED anode
20	LED(-)	Backlight LED cathode

a) pinout

b) graphic LCD layout

c) timing for data write

Figure 7.15: Graphic LCD control signal and timing information.

In Figure 7.16, commands/instructions are provided for the GLCD. Sample code is also provided to get a basic interface constructed between the ATmega164 and the GLCD.

```
//**************************************************************************
//Support software for Hantronix HDM64GS 12 L-4, 128 x 64 graphics LCD
//
//Adapted from Hantronix Application Note and glcd
//**************************************************************************


//**************************************************************************
//Interface notes:  In the code examples provided below, the following
//                  connections were used between the ATmega164 and the
//                  Hamtronix LCD:
```

Instruction	RS	R/W	DB7	DB6	DB5	DB4	DB3	DB2	DB1	DB0	Function
Display ON/OFF	L	L	L	L	H	H	H	H	H	L/H	Display L:off/H: on
Set Address	L	L	L	H	Y addr (0 - 63)						Sets Y address
Set Page (X address)	L	L	H	L	H	H	H	page (0 - 7)			Sets X address
Display Start Line	L	L	H	H	display start line (0 - 63)						Data RAM at top of screen
Status Read	L	H	busy	L	on/off	reset	L	L	L	L	Read status: Busy: ready (L)/in operation(H) on/off: on (L)/off (H) reset: normal (L)/reset (H)
Write Display Data	H	L	write data								Writes data DB[7:0] into display. Y address incremented by 1.
Read Display Data	L	H	read data								Reads data DB[7:0]

Figure 7.16: Graphic LCD commands.

```
//
//                    LCD            ATmega164
//                    /RST(14)       PD5(19)          //reset
//                    R/W-(15)       PD4(18)          //read/write
//                    /CS2(13)       PD3(17)          //chip select 2
//                    /CS1(12)       PD2(16)          //chip select 1
//                    E(17)          PD1(15)          //enable
//                    D/I-(16)       PD0(14)          //data/instruction
//
//*************************************************************************


//function prototypes
unsigned char data_in(void);
void data_out(unsigned char);
void lcd_select_side(unsigned char);
void lcd_data_write(unsigned char);
void lcd_command_write(unsigned char);
unsigned char lcd_data_read(void);
void lcd_set_pixel(unsigned char, unsigned char);
void lcd_wait_busy(void);
void lcd_initialize(void);


//global variables
unsigned char cursor_x, cursor_y;
unsigned char x_address    = 0xB8;
unsigned char y_address    = 0x40;
unsigned char start_line   = 0xC0;
unsigned char display_on   = 0x3F;
unsigned char display_off = 0x3E;
unsigned char busy  = 0x80;
unsigned char right = 0;
unsigned char left  = 1;


//*************************************************************************
```

```c
//function definitions
//****************************************************************************


//****************************************************************************


unsigned char data_in(void)
{
DDRC = 0x00;                              //set PORTC for input
PORTC = 0xFF;
 //pullup resistors
return PINC;                              //read PORTC
}


//****************************************************************************


void data_out(unsigned char data)
{
DDRC = 0xFF;
 //set PORTC for output
PORTC = data;
}


//****************************************************************************


void lcd_select_side (unsigned char side)
{

if(side == right)
  {
   PORTD = PORTD & 0xfd;                  //E=0
   PORTD = PORTD & 0xfe;                  //D/I-=0
   PORTD = PORTD | 0x10;                  //R/W-=1
   PORTD = PORTD & 0xfb;                  //CS1=0
   PORTD = PORTD | 0x08;                  //CS2=1
   lcd_instruction_write(y_address);     //y address
   }
 else
   {
   PORTD = PORTD & 0xfd;                  //E=0
```

```
    PORTD = PORTD & 0xfe;                //D/I-=0
    PORTD = PORTD | 0x10;                //R/W-=1
    PORTD = PORTD | 0x04;                //CS1=1
    PORTD = PORTD & 0xf7;                //CS2=0
    lcd_instruction_write(y_address);    //y address
    }
}


//*************************************************************************

void lcd_data_write(unsigned char data)
{
lcd_wait_busy();

PORTD = PORTD | 0x01;                //D/I-=1
PORTD = PORTD | 0xef;                //R/W-=0
data_out(data);
PORTD = PORTD | 0x02;                //E=1
delay(1);
}


//*************************************************************************


void lcd_instruction_write(unsigned char instruction)
{
lcd_wait_busy();
PORTD = PORTD & 0xfe;                //D/I-=0
PORTD = PORTD | 0xef;                //R/W-=0

data_out(instruction);
PORTD = PORTD | 0x02;                //E=1
delay(1);
}


//*************************************************************************

void lcd_set_pixel(unsigned char x, unsigned char~(y)
```

```
{
unsigned char data_read=0;

lcd_instruction_write(start_line);

if(x<64)
  {
  lcd_select_side(left);
  lcd_instruction_write(x_address + (y / 8));
  lcd_instruction_write(y_address + x);
  data_read = lcd_data_read();
  data_read = lcd_data_read();
  lcd_instruction_write(x_address + (y / 8));
  lcd_instruction_write(y_address + x);
  lcd_data_write(data_read | (1 << (y % 8)));
  }
else
  {
  lcd_select_side(right);
  lcd_instruction_write(x_address + (y / 8));
  lcd_instruction_write(y_address + x - 64);
  data_read = lcd_data_read();
  data_read = lcd_data_read();
  lcd_instruction_write(x_address + (y / 8));
  lcd_instruction_write(y_address + x - 64);
  lcd_data_write(data_read | (1 << (y % 8)));
  }
lcd_instruction_write(start_line);
}

//************************************************************************

void initialize_lcd (void)
{
DDRD = 0xff;                      //set PORTD as output
                                  //use PORTD for LCD control signals

data_out(0x00);                   //data clear
PORTD = PORTD & 0xfe;             //D/I-=0
```

```
PORTD = PORTD | 0xef;              //R/W-=0
PORTD = PORTD & 0xfd;              //E=0
PORTD = PORTD & 0xfb;              //CS1=0
PORTD = PORTD & 0xf7;              //CS2=0
PORTD = PORTD & 0xdf;              //RST-=0
delay(10);
PORTD = PORTD & 0x20;              //RST-=1;
lcd_select_side(left);
lcd_instruction_write(display_off);
lcd_instruction_write(start_line);
lcd_instruction_write(x_address);
lcd_instruction_write(y_address);
lcd_instruction_write(display_on);
lcd_select_side(right);
lcd_instruction_write(display_off);
lcd_instruction_write(start_line);
lcd_instruction_write(x_address);
lcd_instruction_write(y_address);
lcd_instruction_write(display_on);
cls_lcd();
}




//*************************************************************************



void cls_lcd (void)
{
unsigned char  page, column;


for(page=0; page<8; page++)
  {
  lcd_select_side(LEFT);
  lcd_instruction_write(X_ADRESS | page);
  lcd_instruction_write(Y_ADRESS);

  for(column=0; column<128; column++)
```

```
    {
    if(column==64)
{
lcd_select_side(RIGHT);
lcd_instruction_write(X_ADRESS | page);
lcd_instruction_write(Y_ADRESS);
}
    lcd_data_write(0x00);
     }
  }
}


//********************************************************************************
```

7.4.8   HIGH POWER DC DEVICES

A number of direct current devices may be controlled with an electronic switching device such as a MOSFET. Specifically, an N-channel enhancement MOSFET (metal oxide semiconductor field effect transistor) may be used to switch a high current load on and off (such as a motor), using a low current control signal from a microcontroller as shown in Figure 7.17(a). The low current control signal from the microcontroller is connected to the gate of the MOSFET. The MOSFET switches the high current load on and off, consistent with the control signal. The high current load is connected between the load supply and the MOSFET drain. It is important to note that the load supply voltage and the microcontroller supply voltage do not have to be at the same value. When the control signal on the MOSFET gate is logic high, the load current flows from drain to source. When the control signal applied to the gate is logic low, no load current flows. Thus, the high power load is turned on and off by the low power control signal from the microcontroller.

Often the MOSFET is used to control a high power motor load. A motor is a notorious source of noise. To isolate the microcontroller from the motor noise, an optical isolator may be used as an interface as shown in Figure 7.17(b). The link between the control signal from the microcontroller to the high power load is via an optical link contained within a Solid State Relay (SSR). The SSR is properly biased using techniques previously discussed.

7.5   DC SOLENOID CONTROL

The interface circuit for a DC solenoid is provided in Figure 7.18. A solenoid provides a mechanical insertion (or extraction) when asserted. In the interface, an optical isolator is used between the microcontroller and the MOSFET used to activate the solenoid. A reverse biased diode is placed

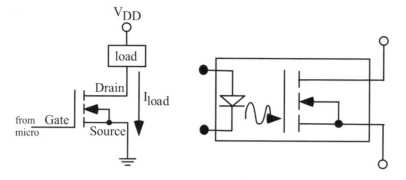

a) N-channel enhance MOSFET   b) solid state relay with optical interface

Figure 7.17: MOSFET circuits.

across the solenoid. Both the solenoid power supply and the MOSFET must have the appropriate voltage and current rating to support the solenoid requirements.

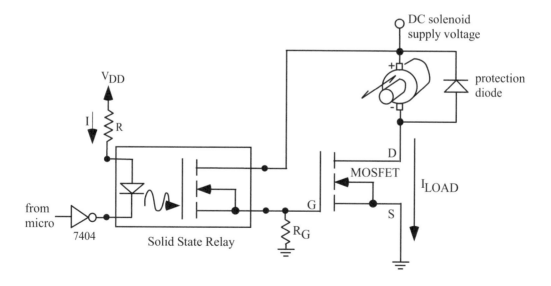

Figure 7.18: Solenoid interface circuit.

7.6   DC MOTOR SPEED AND DIRECTION CONTROL

Often, a microcontroller is used to control a high power motor load. To properly interface the motor to the microcontroller, we must be familiar with the different types of motor technologies. Motor types are illustrated in Figure 7.19.

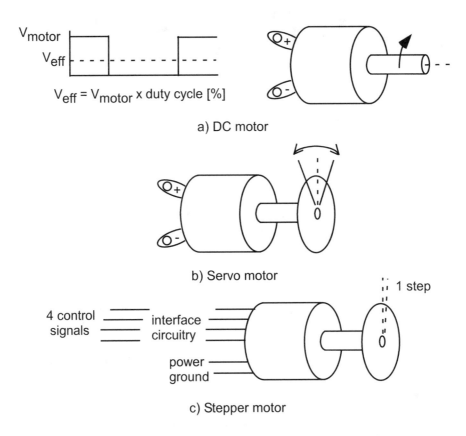

Figure 7.19: Motor types.

- **DC motor:** A DC motor has a positive and negative terminal. When a DC power supply of suitable current rating is applied to the motor, it will rotate. If the polarity of the supply is switched with reference to the motor terminals, the motor will rotate in the opposite direction. The speed of the motor is roughly proportional to the applied voltage up to the rated voltage of the motor.

- **Servo motor:** A servo motor provides a precision angular rotation for an applied pulse width modulation duty cycle. As the duty cycle of the applied signal is varied, the angular displacement of the motor also varies. This type of motor is used to change mechanical positions such as the steering angle of a wheel.

- **Stepper motor:** A stepper motor, as its name implies, provides an incremental step change in rotation (typically 2.5 degree per step) for a step change in control signal sequence. The motor is typically controlled by a two or four wire interface. For the four wire stepper motor, the microcontroller provides a four bit control sequence to rotate the motor clockwise. To turn the motor counterclockwise, the control sequence is reversed. The low power control signals are interfaced to the motor via MOSFETs or power transistors to provide for the proper voltage and current requirements of the pulse sequence.

7.6.1   DC MOTOR OPERATING PARAMETERS

Space does not allow a full discussion of all motor types. We will concentrate on the DC motor. As previously mentioned, the motor speed may be varied by changing the applied voltage. This is difficult to do with a digital control signal. However, PWM control signal techniques discussed earlier may be combined with a MOSFET interface to precisely control the motor speed. The duty cycle of the PWM signal will also be the percentage of the motor supply voltage applied to the motor and hence the percentage of rated full speed at which the motor will rotate. The interface circuit to accomplish this type of control is shown in Figure 7.20. Various portions of this interface circuit have been previously discussed. The resistor R_G, typically 10K ohm, is used to discharge the MOSFET gate when no voltage is applied to the gate. For an inductive load, a reversed biased protection diode must be across the load. The interface circuit shown allows the motor to rotate in a given direction. As previously mentioned, to rotate the motor in the opposite direction the motor polarity must be reversed. This may be accomplished with a high power switching network called an H-bridge specifically designed for this purpose. Reference Pack and Barrett for more information on this topic.

7.6.2   H-BRIDGE DIRECTION CONTROL

For a DC motor to operate in both the clockwise and counter clockwise direction, the polarity of the DC motor supplied must be changed. To operate the motor in the forward direction, the positive battery terminal must be connected to the positive motor terminal while the negative battery terminal must be provided to the negative motor terminal. To reverse the motor direction, the motor supply polarity must be reversed. An H-bridge is a circuit employed to perform this polarity switch. Low power H-bridges (500 mA) come in a convenient dual in line package (e.g., 754110). For higher power motors, a H-bridge may be constructed from discrete components as shown in Figure 7.21. If PWM signals are used to drive the base of the transistors (from microcontroller pins PD4 and PD5), both motor speed and direction may be controlled by the circuit. The transistors used in the circuit must have a current rating sufficient to handle the current requirements of the motor during start and stall conditions.

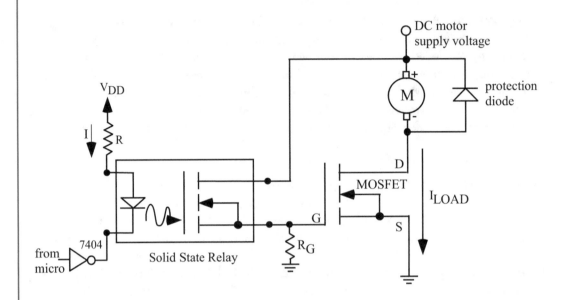

Figure 7.20: DC motor interface.

7.6.3   SERVO MOTOR INTERFACE

The servo motor is used for a precise angular displacement. The displacement is related to the duty cycle of the applied control signal. A servo control circuit and supporting software was provided in Chapter 6.

7.6.4   STEPPER MOTOR CONTROL

Stepper motors are used to provide a discrete angular displacement in response to a control signal step. There are a wide variety of stepper motors including bipolar and unipolar types with different configurations of motor coil wiring. Due to space limitations, we only discuss the unipolar, 5 wire stepper motor. The internal coil configuration for this motor is provided in Figure 7.22b).

Often a wiring diagram is not available for the stepper motor. Based on the wiring config-uration (Reference Figure 7.22b), one can find out the common line for both coils. It will have a resistance that is one-half of all of the other coils. Once the common connection is found, one can connect the stepper motor into the interface circuit. By changing the other connections, one can determine the correct connections for the step sequence.

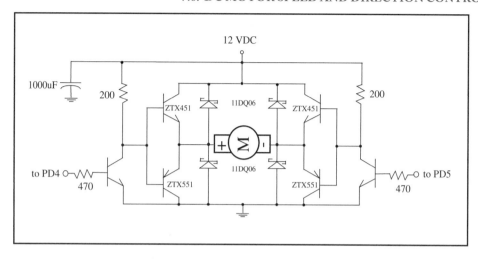

Figure 7.21: H-bridge control circuit. Souce (G. O'Berto).

To rotate the motor either clockwise or counter clockwise, a specific step sequence must be sent to the motor control wires as shown in Figure 7.22c). As shown in Figure 7.22c) the control sequence is transmitted by four pins on the microcontroller. In this example, we use PORTD[7:5].

The microcontroller does not have sufficient capability to drive the motor directly. Therefore, an interface circuit is required as shown in Figure 7.22c). For a unipolar stepper motor, we employ a TIP130 power Darlington transistor to drive each coil of the stepper motor. The speed of motor rotation is determined by how fast the control sequence is completed. The TIP 30 must be powered by a supply that has sufficient capability for the stepper motor coils.

Example: An ATmega324 has been connected to a JRP 42BYG016 unipolar, 1.8 degree per step, 12 VDC at 160 mA stepper motor. The interface circuit is shown in Figure 7.23. PORTD pins 7 to 4 are used to provide the step sequence. A one second delay is used between the steps to control motor speed. Pushbutton switches are used on PORTB[1:0] to assert CW and CCW stepper motion. An interface circuit consisting of four TIP130 transistors are used between the microcontroller and the stepper motor to boost the voltage and current of the control signals. Code to provide the step sequence is shown below.

Provided below is a basic function to rotate the stepper motor in the forward or reverse direction.

```
//****************************************************************************
//target controller: ATMEL ATmega324
//
//ATMEL AVR ATmega324PV Controller Pin Assignments
//Chip Port Function I/O Source/Dest Asserted Notes
```

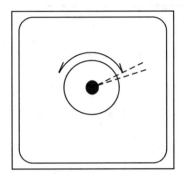

a) a stepper motor rotates a fixed angle per step

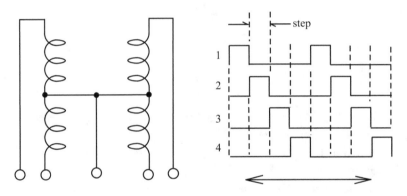

b) coil configuration and step sequence

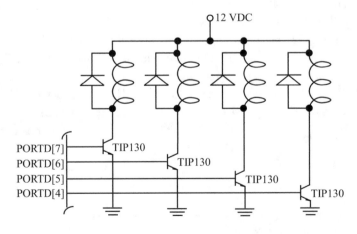

c) stepper motor interface circuit

Figure 7.22: Unipolar stepper motor control circuit.

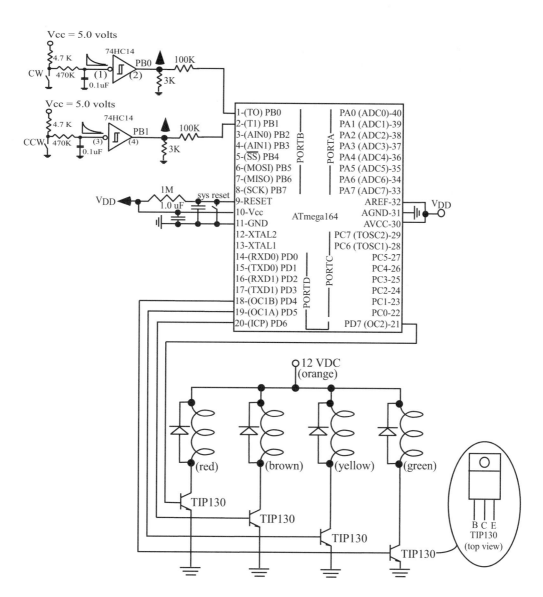

c) stepper motor interface circuit

Figure 7.23: Unipolar stepper motor control circuit.

```
//PORTB:
//Pin 1 PB0 to active high RC debounced switch - CW
//Pin 2 PB1 to active high RC debounced switch - CCW
//Pin 9 Reset - 1M resistor to Vcc, tact switch to ground, 1.0 uF to ground
//Pin 10 Vcc - 1.0 uF to ground
//Pin 11 Gnd
//Pin 12 ZTT-10.00MT ceramic resonator connection
//Pin 13 ZTT-10.00MT ceramic resonator connection
//Pin 18 PD4 - to servo control input
//Pin 30 AVcc to Vcc
//Pin 31 AGnd to Ground
//Pin 32 ARef to Vcc
//****************************************************************************
//include files**************************************************************

//ATMEL register definitions for ATmega324
#include<iom324pv.h>
#include<macros.h>
                                              //interrupt handler definition
#pragma interrupt_handler timer0_interrupt_isr:19



//function prototypes********************************************************
void initialize_ports(void);           //initializes ports
void power_on_reset(void);              //returns system to startup state
void read_new_input(void);
//used to read input change on PORTB
void init_timer0_ovf_interrupt(void);
//used to initialize timer0 overflow
void timer0_interrupt_isr(void);
void delay(unsigned int);



//main program***************************************************************
//The main program checks PORTB for user input activity. If new activity
//is found, the program responds.

//global variables
unsigned char   old_PORTB = 0x08; //present value of PORTB
```

```
unsigned char    new_PORTB;        //new values of PORTB
unsigned int     input_delay;      //delay counter - increment via Timer0
                                   //overflow interrupt


void main(void)
{
initialize_ports();               //return LED configuration to default
init_timer0_ovf_interrupt();      //used to initialize timer0 overflow

while(1)
  {
  _StackCheck();                  //check for stack overflow
  read_new_input();               //read input status changes on PORTB
  }
}//end main

//Function definitions
//****************************************************************************
//initialize_ports: provides initial configuration for I/O ports
//****************************************************************************

void initialize_ports(void)
{
//PORTA
DDRA=0xff;                        //PORTA[7:0] output
PORTA=0x00;                       //turn off pull ups

//PORTB
DDRB=0xfc;                        //PORTB[7-2] output, PORTB[1:0] input
PORTB=0x00;                       //disable PORTB pull-up resistors

//PORTC
DDRC=0xff;                        //set PORTC[7-0] as output
PORTC=0x00;                       //init low

//PORTD
DDRD=0xff;                        //set PORTD[7-0] as output
PORTD=0x00;                       //initialize low
}
```

```c
//****************************************************************************
//read_new_input: functions polls PORTB for a change in status. If status
//change has occurred, appropriate function for status change is called
//Pin 1 PB0 to active high RC  debounced switch - CW
//Pin 2 PB1 to active high RC debounced switch - CCW
//****************************************************************************

void read_new_input(void)
{
new_PORTB = (PINB & 0x03);
if(new_PORTB != old_PORTB){
  switch(new_PORTB){                  //process change in PORTB input

    case 0x01:                        //CW
      while((PINB & 0x03)==0x01)
        {
        PORTD = 0x80;
        delay(15);                    //delay 1s
        PORTD = 0x00;
        delay(1);                     //delay 65 ms

        PORTD = 0x40;
        delay(15);
        PORTD = 0x00;
        delay(1);

        PORTD = 0x20;
        delay(15);
        PORTD = 0x00;
        delay(1);

        PORTD = 0x10;
        delay(15);
        PORTD = 0x00;
        delay(1);
        }
    break;
```

```
  case 0x02:                       //CCW
    while((PINB & 0x03)==0x02)
      {
      PORTD = 0x10;
      delay(15);
      PORTD = 0x00;
      delay(1);

      PORTD = 0x20;
      delay(15);
      PORTD = 0x00;
      delay(1);

      PORTD = 0x40;
      delay(15);
      PORTD = 0x00;
      delay(1);

      PORTD = 0x80;
      delay(15);
      PORTD = 0x00;
      delay(1);
 }
      break;

    default:;                      //all other cases
      }                            //end switch(new_PORTB)
  }                                //end if new_PORTB
  old_PORTB=new_PORTB;             //update PORTB
}


//*****************************************************************************
//int_timer0_ovf_interrupt(): The Timer0 overflow interrupt is being
//employed as a time base for a master timer for this project. The internal
//oscillator of 8 MHz is divided internally by 8 to provide a 1 MHz time base
//and is divided by 256.  The 8-bit Timer0
register (TCNT0) overflows every
//256 counts or every 65.5 ms.
//*****************************************************************************
```

```
void init_timer0_ovf_interrupt(void)
{
TCCR0B = 0x04; //divide timer0 timebase
by 256, overflow occurs every 65.5ms
TIMSK0 = 0x01; //enable timer0 overflow interrupt
asm("SEI");    //enable global interrupt
}


//*************************************************************************
//timer0_interrupt_isr:
//Note: Timer overflow 0 is cleared by hardware when executing the
//corresponding interrupt handling vector.
//*************************************************************************

void timer0_interrupt_isr(void)
{
input_delay++;                        //input delay processing
}


//*************************************************************************
//void delay(unsigned int number_of_65_5ms_interrupts)
//this generic delay function provides the specified delay as the number
//of 65.5 ms "clock ticks" from the Timer0 interrupt.
//Note: this function is only valid when using a 1 MHz crystal or ceramic
//      resonator
//*************************************************************************

void delay(unsigned int number_of_65_5ms_interrupts)
{
TCNT0 = 0x00;                     //reset timer0
input_delay = 0;
while(input_delay <= number_of_65_5ms_interrupts)
  {
  ;
  }
}


//*************************************************************************
```

7.6.5   AC DEVICES

In a similar manner, a high power alternating current (AC) load may be switched on and off using a low power control signal from the microcontroller. In this case, a Solid State Relay is used as the switching device. Solid state relays are available to switch a high power DC or AC load (Crydom). For example, the Crydom 558-CX240D5R is a printed circuit board mounted, air cooled, single pole single throw (SPST), normally open (NO) solid state relay. It requires a DC control voltage of 3-15 VDC at 15 mA. However, this small microcontroller compatible DC control signal is used to switch 12-280 VAC loads rated from 0.06 to 5 amps (Crydom) as shown in Figure 7.24.

To vary the direction of an AC motor you must use a bi-directional AC motor. A bi-directional motor is equipped with three terminals: common, clockwise, and counterclockwise. To turn the motor clockwise, an AC source is applied to the common and clockwise connections. In like manner, to turn the motor counterclockwise, an AC source is applied to the common and counterclockwise connections. This may be accomplished using two of the Crydom SSRs.

7.7   INTERFACING TO MISCELLANEOUS DEVICES

In this section we provide a pot pourri of interface circuits to connect a microcontroller to a wide variety of peripheral devices.

7.7.1   SONALERTS, BEEPERS, BUZZERS

In Figure 7.25, we provide several circuits used to interface a microcontroller to a buzzer, beeper or other types of annunciator indexannunciator devices such as a sonalert. It is important that the interface transistor and the supply voltage are matched to the requirements of the sound producing device.

7.7.2   VIBRATING MOTOR

A vibrating motor is often used to gain one's attention as in a cell phone. These motors are typically rated at 3 VDC and a high current. The interface circuit shown in Figure 7.26 is used to drive the low voltage motor. Note that the control signal provided to the transistor base is 5 VDC. To step the motor supply voltage down to the motor voltage of 3 VDC, two 1N4001 silicon rectifier diodes are used in series. These diodes provide approximately 1.4 to 1.6 VDC voltage drop. Another 1N4001 diode is reversed biased across the motor and the series diode string. The motor may be turned on and off with a 5 VDC control signal from the microcontroller, or a PWM signal may be used to control motor speed.

7.7.3   DC FAN

An interface circuit to control a DC fan is provided in Figure 7.27. An optical solid state relay is used to isolate the motor from the microcontroller. This provides noise isolation for the microcontroller. A reverse biased diode is placed across the DC motor.

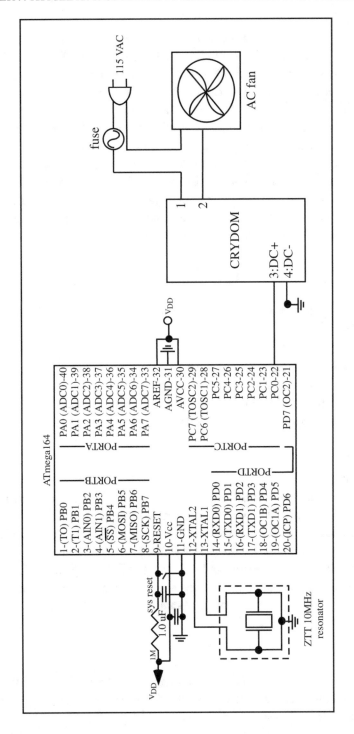

Figure 7.24: AC motor control circuit.

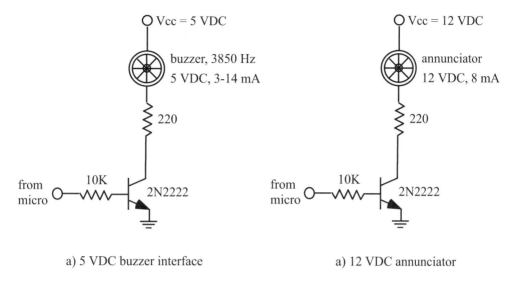

Figure 7.25: Sonalert, beepers, buzzers.

7.8 SUMMARY

In this chapter, we discussed the voltage and current operating parameters for the Atmel HC CMOS type microcontroller. We discussed how this information may be applied to properly design an interface for common input and output circuits. It must be emphasized, a properly designed interface allows the microcontroller to operate properly within its parameter envelope. If due to a poor interface design, a microcontroller is used outside its prescribed operating parameter values, spurious and incorrect logic values will result. We provided interface information for a wide range of input and output devices. We also discussed the concept of interfacing a motor to a microcontroller using PWM techniques coupled with high power MOSFET or SSR switching devices.

7.9 CHAPTER PROBLEMS

7.1. What will happen if a microcontroller is used outside of its prescribed operating envelope?

7.2. Discuss the difference between the terms "sink" and "source" as related to current loading of a microcontroller.

7.3. Can an LED with a series limiting resistor be directly driven by the Atmel microcontroller? Explain.

7.4. In your own words, provide a brief description of each of the microcontroller electrical parameters.

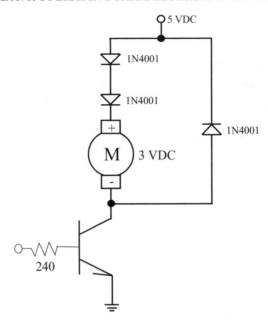

Figure 7.26: Controlling a low voltage motor.

7.5. What is switch bounce? Describe two techniques to minimize switch bounce.

7.6. Describe a method of debouncing a keypad.

7.7. What is the difference between an incremental encoder and an absolute encoder? Describe applications for each type.

7.8. What must be the current rating of the 2N2222 and 2N2907 transistors used in the tri-state LED circuit? Support your answer.

7.9. Draw the circuit for a six character seven segment display. Fully specify all components. Write a program to display "ATmega16. "

7.10. Repeat the question above for a dot matrix display.

7.11. Repeat the question above for a LCD display.

7.12. What is the difference between a unipolar and bipolar stepper motor?

7.13. What controls the speed of rotation of a stepper motor?

7.14. A stepper motor provides and angular displacement of 1.8 degrees per step. How can this resolution be improved?

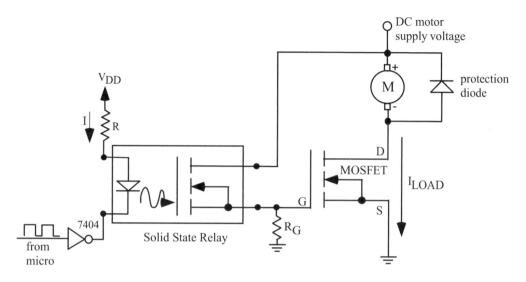

Figure 7.27: DC fan control.

7.15. Write a function to convert an ASCII numeral representation (0 to (9) to a seven segment display.

7.16. Why is an interface required between a microcontroller and a stepper motor?

7.17. Construct UML activity diagrams for the GLCD functions provided in the chapter.

REFERENCES

D. Pack and S. BarrettS (2002) 68HC12 Microcontroller: Theory and Applications. Prentice-Hall Incorporated, Upper Saddle River, NJ.

S. Barrett and D. Pack (2004) Embedded Systems Design with the 68HC12 and HCS12. Prentice-Hall Incorporated, Upper Saddle River, NJ.

Crydom Corporation, 2320 Paseo de las Americas, Suite 201, San Diego, CA (www.crydom.com).

Sick/Stegmann Incorporated, Dayton, OH, (www.stegmann.com).

Images Company, 39 Seneca Loop, Staten Island, NY 10314.

Atmel 8-bit AVR Microcontroller with 16/32/64K Bytes In-System Programmable Flash, ATmega164P/V, ATmega324P/V, 644P/V data sheet: 8011I-AVR-05/08, Atmel Corporation, 2325 Orchard Parkway, San Jose, CA 95131.

S. Barrett and D. Pack (2006) Microcontrollers Fundamentals for Engineers and Scientists. Morgan and Claypool Publishers.

National Semiconductor, *LM34/LM34A/LM34C/LM34CA/LM34D Precision Fahrenheit Temperature Sensor*, 1995.

Hantronix Inc. *Hantronix Product Specification HDM64Gs12L-4 128 x 64 Graphics LCD Display Module*, Hantronix Inc., 10080 Bubb Rd. Cupertino, CA 95014, 2003.

CHAPTER 8

System Level Design

Objectives: After reading this chapter, the reader should be able to

- Design an embedded system requiring a variety of microcontroller subsystems and input and output devices.

- Design circuits to interface the microcontroller with required system input and output devices.

- Employ a variety of tools to design embedded systems.

8.1 OVERVIEW

In this chapter, we design three different microcontroller-based embedded systems to illustrate concepts presented throughout the text. We have chosen these systems to expose the reader to a wide variety of requirements, peripheral devices, and interface techniques for microcontroller-based embedded systems. We provide basic designs for the three systems and challenge the reader to extend the designs with additional features. The three systems are:

- a weather station,

- a motor speed control circuit, and

- an autonomous maze navigating robot.

   For each system we provide the following:

- a system description,

- system requirements,

- a structure chart,

- a system circuit diagram,

- UML activity diagrams, and

- the associated microcontroller code.

8.2   WEATHER STATION

In this project, we design a weather station to sense wind direction and to measure ambient temperature. The measure temperature values are displayed on an LCD in Fahrenheit. The wind direction is displayed on LEDs arranged in a circular pattern. The wind direction and temperature are transmitted serially to an external device.

8.2.1   REQUIREMENTS
The requirements for this system include:

- Sense wind direction and measure ambient temperature.

- Display on an LCD.

- Display measured temperature in Fahrenheit on an LCD.

- Display wind direction on LEDs arranged in a circular pattern.

- Transmit serially wind direction and temperature data.

8.2.2   STRUCTURE CHART
To begin the design process, a structure chart is used to partition the system into definable subsystems. We employ a top-down design/bottom-up implementation approach. The structure chart for the weather station is shown in Figure 8.1. The three main microcontroller subsystems needed for this project are the USART for serial communication, the ADC system to convert the analog voltages from the LM34 temperature sensor and a weather vane into digital signals, and the wind direction display. This display consists of a 74154, 4-to-16 decoder and 16 individual LEDs to display wind direction. The system is partitioned until the lowest level of the structure chart contains "doable" pieces of hardware components or software functions. Data flow is shown on the structure chart as directed arrows.

8.2.3   CIRCUIT DIAGRAM
The circuit diagram for the weather station is shown in Figure 8.2. The weather station is equipped with two input sensors: the LM34 to measure temperature and the weather vane to measure wind direction. Both of the sensors provide an analog output that is fed to PORTA[0] and PORTA[1]. The LM34 provides 10 mV output per degree Fahrenheit. The weather vane provides 0 to 5 VDC for 360 degrees of vane rotation. The weather vane must be oriented to a known direction with the output voltage at this direction noted. We assume that 0 VDC corresponds to North and the voltage increases as the vane rotates clockwise to the East. The vane output voltage continues to increase until North is again reached at 5 VDC and then rolls over back to zero volts. All other directions are derived from this reference point.

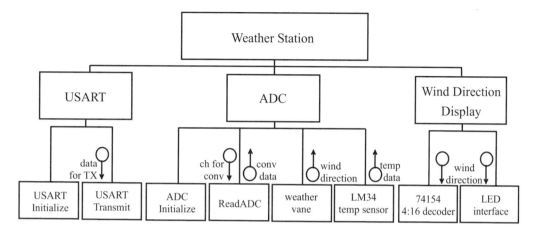

Figure 8.1: Weather station structure chart.

An LCD is connected to PORTC for data and PORTD[7:6] for the enable and command/data control lines. There are 16 different LEDs for the wind speed indicator. Rather than use 16 microcontroller pins, the binary value of the LED for illumination should be sent to the 74154 4-to-16 decoder. The decoder provides a "one cold" output as determined by the binary code provided on PORTA[7:4]. For example, when A_{16} is sent to the 74154 input, output $/Y10$ is asserted low, while all other outputs remain at logic high. The 74154 is from the standard TTL family. It has sufficient current sink capability ($I_{OL} = 16\,mA$) to meet the current requirements of an LED ($V_f = 1.5\,VDC$, $I_f = 15\,mA$).

8.2.4   UML ACTIVITY DIAGRAMS
The UML activity diagram for the main program is shown in Figure 8.3. After initializing the subsystems, the program enters a continuous loop where temperature and wind direction are sensed and displayed on the LCD and the LED display. The sensed values are then transmitted via the USART. The system then executes a delay routine to configure how often the temperature and wind direction parameters should be updated. We leave the construction of the individual UML activity diagrams for each function as an end of chapter exercise.

8.2.5   MICROCONTROLLER CODE
```
//include file*********************************************************
#include <iom164pv.h>


//function prototypes**************************************************
void initialize_ports(void);
```

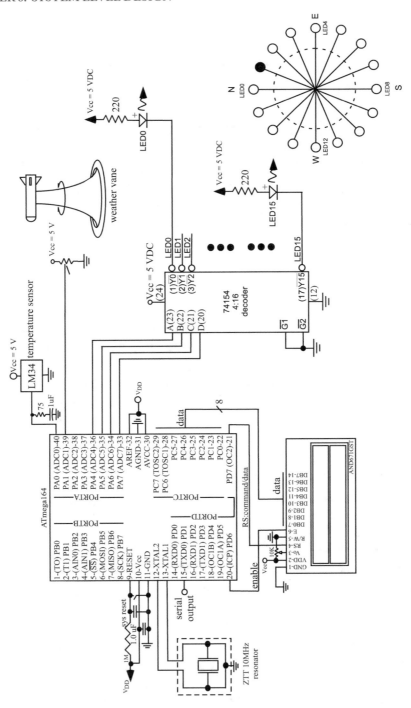

Figure 8.2: Circuit diagram for weather station.

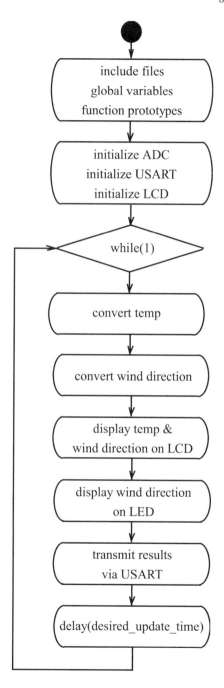

Figure 8.3: Weather station UML activity diagram.

```c
void initialize_ADC(void);
void temperature_to_LCD(unsigned int ADCValue);
unsigned int readADC(unsigned char);
void LCD_init(void);
void putChar(unsigned char);
void putcommand(unsigned char);
void display_data(void);
void InitUSART_ch1(void);
void USART_TX_ch1(unsigned char data);
void convert_wind_direction(unsigned int);
void delay(unsigned int number_of_6_55ms_interrupts);
void init_timer0_ovf_interrupt(void);
void timer0_interrupt_isr(void);


                                            //interrupt handler definition
#pragma interrupt_handler timer0_interrupt_isr:19


    //door profile data



//Global Variables*********************************************************
unsigned int temperature, wind_direction;
unsigned int binary_weighted_voltage_low, binary_weighted_voltage_high;
unsigned char dir_tx_data;

void main(void)
{
initialize_ports();
initialize_ADC();
void InitUSART_ch1();
LCD_init();
init_timer0_ovf_interrupt();

while(1)
  {
  //temperature data: read -> display -> transmit
  temperature = readADC(0x00);              //Read temp from LM34
  temperature_to_LCD (temperature);
    //Convert and display temp on LCD
```

```
  USART_TX_ch1((unsigned char)(binary_weighted_voltage_low));
  USART_TX_ch1((unsigned char)(binary_weighted_voltage_high >>8));

  //wind direction data: read -> display -> transmit
  wind_direction = readADC(0x01);          //Read wind direction
  convert_wind_direction(wind_direction);
//Convert wind direction -> transmit
  USART_TX_ch1((unsigned char)(binary_weighted_voltage_low));
  USART_TX_ch1((unsigned char)(binary_weighted_voltage_high >>8));

  //delay 15 minutes
  delay(2307):
  }
}


//****************************************************************

void initialize_ports()
{
DDRD = 0xFB;
DDRC = 0xFF;
DDRB = 0xFF;
}


//****************************************************************

void initialize_ADC()
{
ADMUX = 0;                               //Select channel 0

 //Enable ADC and set module enable ADC
ADCSRA = 0xC3;                           //Set module prescalar to 8
while(!(ADCSRA & 0x10));                 //Wait until conversion is ready
ADCSRA |= 0x10;                          //Clear conversion ready flag
}


//****************************************************************

unsigned int readADC(unsigned char channel)
```

```c
{
unsigned int binary_weighted_voltage, binary_weighted_voltage_low;
unsigned int binary_weighted_voltage_high;//weighted binary voltage

ADMUX = channel;                          //Select channel
ADCSRA |= 0x43;                           //Start conversion
                                          //Set ADC module prescalar
                                          //to 8 critical for
                                          //accurate ADC results
while (!(ADCSRA & 0x10));                 //Check if conversion is ready
ADCSRA |= 0x10;
 //Clear conv rdy flag - set the bit
binary_weighted_voltage_low = ADCL;
 //Read 8 low bits first (important)

 //Read 2 high bits, multiply by 256
binary_weighted_voltage_high = ((unsigned int)(ADCH << 8));
binary_weighted_voltage = binary_weighted_voltage_low
+ binary_weighted_voltage_high;
return binary_weighted_voltage;           //ADCH:ADCL

}

//**************************************************************************
//LCD_Init: initialization for AND671GST 1x16 character display LCD
//LCD configured as two 8 character lines in a 1x16 array
//LCD data bus (pin 14-pin7) ATMEL ATmega16: PORTC
//LCD RS (pin~(4) ATMEL ATmega16: PORTD[7]
//LCD E (pin~(6) ATMEL ATmega16: PORTD[6]
//**************************************************************************

void LCD_init(void)
{
delay(1);
delay(1);
delay(1);
                                          //Output command string to
                                          //Initialize LCD
putcommand(0x38);                         //Function set 8-bit
```

```
delay(1);
putcommand(0x38);                       //Function set 8-bit
delay(1);
putcommand(0x38);                       //Function set 8-bit
putcommand(0x38);                       //One line, 5x7 char
putcommand(0x0E);                       //Display on
putcommand(0x01);                       //Display clear-1.64 ms
putcommand(0x06);                       //Entry mode set
putcommand(0x00);                       //Clear display, cursor at home
putcommand(0x00);                       //Clear display, cursor at home
}



//***********************************************************************


void putChar(unsigned char~(c)
{
DDRC = 0xff;                            //Set PORTC as output
DDRD = DDRD|0xC0;                       //Make PORTD[7:6] output
PORTC = c;
PORTD = PORTD|0x80;                     //RS=1
PORTD = PORTD|0x40;                     //E=1
PORTD = PORTD&0xbf;                     //E=0
delay(1);
}


//***********************************************************************


void putcommand(unsigned char~(d)
{
DDRC = 0xff;                            //Set PORTC as output
DDRD = DDRD|0xC0;                       //Make PORTD[7:6] output
PORTD = PORTD&0x7f;                     //RS=0
PORTC = d;
PORTD = PORTD|0x40; //E=1
PORTD = PORTD&0xbf; //E=0
delay(1);
}
```

```
//**********************************************************************

void temperature_to_LCD(unsigned int ADCValue)
{
float voltage,temperature;
unsigned int tens, ones, tenths;

voltage = (float)ADCValue*5.0/1024.0;

temperature = voltage*100;


tens = (unsigned int)(temperature/10);
ones = (unsigned int)(temperature-(float)tens*10);
tenths = (unsigned int)(((temperature-(float)tens*10)-(float)ones)*10);

putcommand(0x01);                       //Cursor home
putcommand(0x80);                       //DD RAM location 1 - line 1
putChar((unsigned char)(tens)+48);
putChar((unsigned char)(ones)+48);
putChar('.');
putChar((unsigned char)(tenths)+48);
putChar('F');
}


//**********************************************************************

void convert_wind_direction(unsigned int wind_dir_int)
{
float wind_dir_float;
                                        //Convert wind direction to float
wind_dir_float = ((float)wind_dir_int)/1024.0) * 5;

//N - LED0
if((wind_dir_float <= 0.15625)||(wind_dir_float > 4.84375))
   {
   putcommand(0x01);                    //Cursor to home
   putcommand(0xc0);                    //DD RAM location 1 - line 2
   putchar('N');                        //LCD displays: N
```

```
    PORTA = 0x00;                          //Illuminate LED 0

    }

//NNE - LED1
if((wind_dir_float > 0.15625)||(wind_dir_float <= 0.46875))
    {
    putcommand(0x01);                      //Cursor to home
    putcommand(0xc0);                      //DD RAM location 1 - line 2
    putchar('N');                          //LCD displays: NNE
    putchar('N');
    putchar('E');
    PORTA = 0x10;                          //Illuminate LED 1
    }

//NE - LED2
if((wind_dir_float > 0.46875)||(wind_dir_float <= 0.78125))
    {
    putcommand(0x01);                      //Cursor to home
    putcommand(0xc0);                      //DD RAM location 1 - line 2
    putchar('N');                          //LCD displays: NE
    putchar('E');
    PORTA = 0x20;                          //Illuminate LED 2
    }

//ENE - LED3
if((wind_dir_float > 0.78125)||(wind_dir_float <= 1.09375))
    {
    putcommand(0x01);                      //Cursor to home
    putcommand(0xc0);                      //DD RAM location 1 - line 2
    putchar('E');                          //LCD displays: NNE
    putchar('N');
    putchar('E');
    PORTA = 0x30;                          //Illuminate LED 3

    }

//E - LED4
if((wind_dir_float > 1.09375)||(wind_dir_float <= 1.40625))
```

```
     {
     putcommand(0x01);                      //Cursor to home
     putcommand(0xc0);                      //DD RAM location 1 - line 2
     putchar('E');                          //LCD displays: E
     PORTA = 0x40;                          //Illuminate LED 4
     }

//ESE - LED5
if((wind_dir_float > 1.40625)||(wind_dir_float <= 1.71875))
     {
     putcommand(0x01);                      //Cursor to home
     putcommand(0xc0);                      //DD RAM location 1 - line 2
     putchar('E');                          //LCD displays: ESE
     putchar('S');
     putchar('E');
     PORTA = 0x50;                          //Illuminate LED 5


     }



//SE - LED6
if((wind_dir_float > 1.71875)||(wind_dir_float <= 2.03125))
     {
     putcommand(0x01);                      //Cursor to home
     putcommand(0xc0);                      //DD RAM location 1 - line 2
     putchar('S');                          //LCD displays: SE
     putchar('E');
     PORTA = 0x60;                          //Illuminate LED 6
     }

//SSE - LED7
if((wind_dir_float > 2.03125)||(wind_dir_float <= 2.34875))
     {
     putcommand(0x01);                      //Cursor to home
     putcommand(0xc0);                      //DD RAM location 1 - line 2
     putchar('S');                          //LCD displays: SSE
     putchar('S');
     putchar('E');
     PORTA = 0x70;                          //Illuminate LED 7
```

```
    }

//S - LED8
if((wind_dir_float > 2.34875)||(wind_dir_float <= 2.65625))
    {
    putcommand(0x01);                       //Cursor to home
    putcommand(0xc0);                       //DD RAM location 1 - line 2
    putchar('S');                           //LCD displays: S

    PORTA = 0x80;                           //Illuminate LED 8

    }

//SSW - LED9
if((wind_dir_float > 2.65625)||(wind_dir_float <= 2.96875))
    {
    putcommand(0x01);                       //Cursor to home
    putcommand(0xc0);                       //DD RAM location 1 - line 2
    putchar('S');                           //LCD displays: SSW
    putchar('S');
    putchar('W');
    PORTA = 0x90;                           //Illuminate LED 9
    }

//SW - LED10 (A)
if((wind_dir_float > 2.96875)||(wind_dir_float <= 3.28125))
    {
    putcommand(0x01);                       //Cursor to home
    putcommand(0xc0);                       //DD RAM location 1 - line 2
    putchar('S');                           //LCD displays: SW
    putchar('W');
    PORTA = 0xa0;                           //Illuminate LED 10 (A)
    }


//WSW - LED11 (B)
if((wind_dir_float > 3.28125)||(wind_dir_float <= 3.59375))
    {
    putcommand(0x01);                       //Cursor to home
```

```
      putcommand(0xc0);                      //DD RAM location 1 - line 2
      putchar('W');                          //LCD displays: WSW
      putchar('S');
      putchar('W');
      PORTA = 0xb0;                          //Illuminate LED 11 (B)
      }

//W - LED12 (C)
if((wind_dir_float > 3.59375)||(wind_dir_float <= 3.90625))
      {
      putcommand(0x01);                      //Cursor to home
      putcommand(0xc0);                      //DD RAM location 1 - line 2
      putchar('W');                          //LCD displays: W
      PORTA = 0xc0;                          //Illuminate LED 12 (C)
      }

//WNW - LED13 (D)


if((wind_dir_float > 3.90625)||(wind_dir_float <= 4.21875))
      {
      putcommand(0x01);                      //Cursor to home
      putcommand(0xc0);                      //DD RAM location 1 - line 2
      putchar('W');                          //LCD displays: WNW
      putchar('N');
      putchar('W');
      PORTA = 0xd0;                          //Illuminate LED 13 (D)
      }

//NW - LED14 (E)
if((wind_dir_float > 4.21875)||(wind_dir_float <= 4.53125))
      {
      putcommand(0x01);                      //Cursor to home
      putcommand(0xc0);                      //DD RAM location 1 - line 2
      putchar('N');                          //LCD displays: NW
      putchar('W');
      PORTA = 0xe0;                          //Illuminate LED 14 (E)


      }
```

```
//NNW - LED15(F)
if((wind_dir_float > 4.53125)||(wind_dir_float < 4.84375))
   {
   putcommand(0x01);                  //Cursor to home
   putcommand(0xc0);                  //DD RAM location 1 - line 2
   putchar('N');                      //LCD displays: NNW
   putchar('N');
   putchar('W');
   PORTA = 0xf0;                      //Illuminate LED 15 (F)


   }
}


//****************************************************************************

void InitUSART_ch1(void)
{
//USART Channel 1 initialization
//System operating frequency: 10 MHz
// Comm Parameters: 8 bit Data, 1 stop, No Parity
// USART Receiver: Off
// USART Trasmitter: On
// USART Mode: Asynchronous
// USART Baud Rate: 9600

UCSR1A=0x00;
UCSR1B=0x18;                          //RX on, TX on
UCSR1C=0x06;                          //1 stop bit, No parity
UBRR1H=0x00;
UBRR1L=0x40;
}


//****************************************************************************

void USART_TX_ch1(unsigned char data)
{
                                      //Set USART Data Register
                                      //data register empty?
```

```c
while(!(UCSR1A & 0x20));
 UDR1= data;                             //Sets the value in ADCH to the
                                         //value in the USART Data Register

}


//********************************************************************
unsigned char USART_RX_ch1(void)
{
unsigned char rx_data;

 //Checks to see if receive is complete
while(!(UCSR1A & 0x80));
rx_data=UDR1;                           // Returns data
return rx_data;
}


//********************************************************************
//int_timer0_ovf_interrupt(): The Timer0 overflow interrupt is being
//employed as a time base for a master timer for this project. The ceramic
//resonator operating at 10 MHz is divided by 256.  The 8-bit Timer0
//register (TCNT0) overflows every 256 counts or every 6.55 ms.
//********************************************************************

void init_timer0_ovf_interrupt(void)
{
TCCR0B = 0x04; //divide timer0 timebase
by 256, overflow occurs every 6.55ms
TIMSK0 = 0x01; //enable timer0 overflow interrupt
asm("SEI");    //enable global interrupt
}


//********************************************************************
//timer0_interrupt_isr:
//Note: Timer overflow 0 is cleared by hardware when executing the
//corresponding interrupt handling vector.
//********************************************************************


void timer0_interrupt_isr(void)
```

```
{
input_delay++;                              //input delay processing
}


//**********************************************************************
//delay(unsigned int num_of_6_55ms_interrupts): This generic delay function
//provides the specified delay as the number of 6.55 ms "clock ticks" from
//the Timer0 interrupt.
//Note: This function is only valid when using a 10 MHz crystal or ceramic
//      resonator
//**********************************************************************

void delay(unsigned int number_of_6_55ms_interrupts)
{
TCNT0 = 0x00;                               //reset timer0
input_delay = 0;
while(input_delay <= number_of_6_55ms_interrupts)
  {
  ;

  }
}


//**********************************************************************
```

8.3   MOTOR SPEED CONTROL

In this project, we will control the speed of a 24 VDC, 1500 RPM motor using an external potentiometer. The motor is equipped with an optical tachometer, which produces three channels of information. Two of the channels output quadrature related (90 degrees out of phase with one another) 0.5 V peak sinusoidal signals. The third channel provides a single pulse index signal for every motor revolution as shown in Figure 8.4.

8.4   CIRCUIT DIAGRAM

We employ the pulse width modulation (PWM) system of the Atmel ATmega164 to set the motor speed as determined by the potentiometer setting. A potentiometer setting of 0 VDC equates to a 50% duty cycle; whereas, a 5 VDC setting corresponds to a 100% duty cycle. The motor speed and duty cycle will be displayed on an LCD as shown in Figure 8.5.

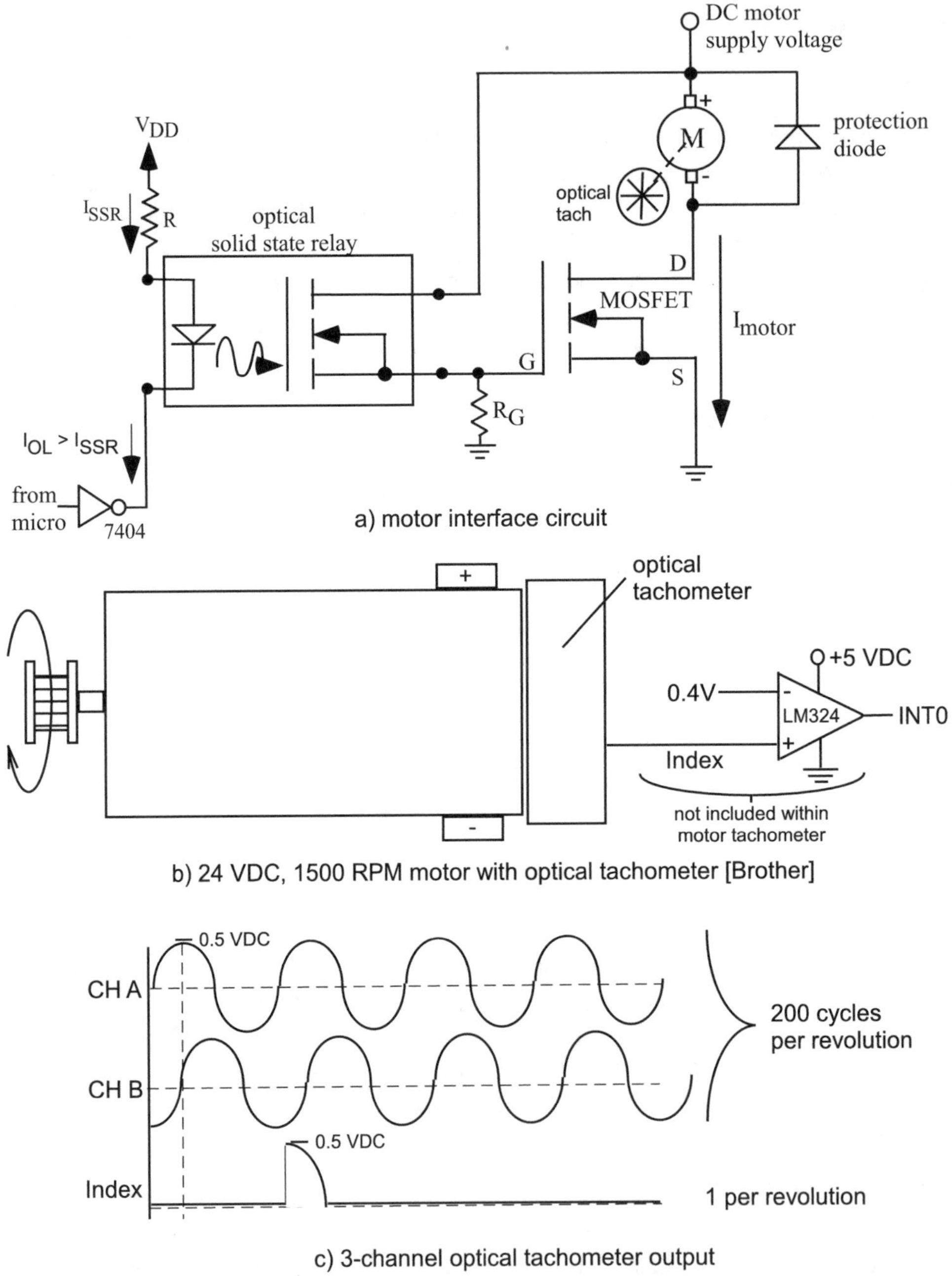

a) motor interface circuit

b) 24 VDC, 1500 RPM motor with optical tachometer [Brother]

c) 3-channel optical tachometer output

Figure 8.4: 24 VDC, 1500 RPM motor equipped with a 3 channel optical encoder.

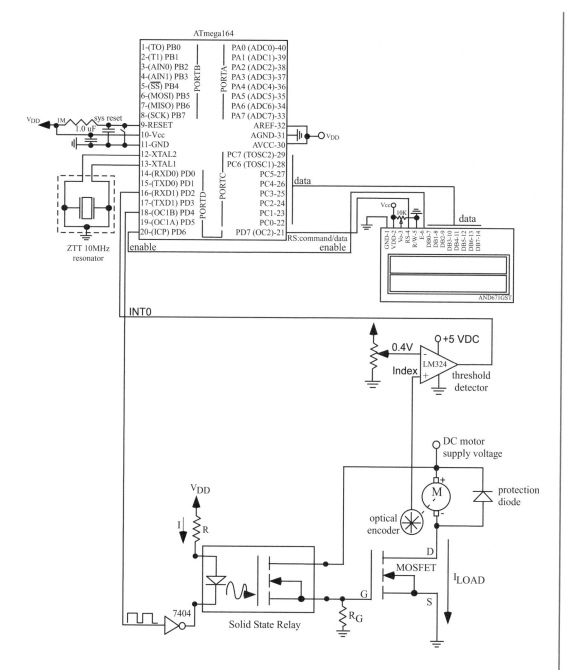

Figure 8.5: Circuit diagram for a 24 VDC, 1500 RPM motor equipped with a 3 channel optical encoder.

In this application, the PWM baseline frequency may be set to a specific frequency. The duty cycle will be varied to adjust the effective voltage delivered to the motor. For example, a 50% duty cycle will deliver an effective value of 50% of the DC motor supply voltage to the motor.

The microcontroller is not directly connected to the motor. The PWM control signal from OC1B (pin 18) is fed to the motor through an optical solid state relay (SSR) as shown in Figure 8.5. This isolates the microcontroller from the noise of the motor. The output signal from SSR is fed to the MOSFET which converts the low-level control signal to voltage and current levels required by the motor.

Motor speed is monitored via the optical encoder connected to the motor shaft. The index output of the motor provides a pulse for each rotation of the motor. The signal is converted to a TTL compatible signal via the LM324 threshold detector. The output from this stage is fed back to INTO to trigger an external interrupt. An interrupt service routine captures the time since the last interrupt. This information is used to speed up or slow down the motor to maintain a constant speed.

8.4.1   REQUIREMENTS

- Generate a 1 kHz PWM signal.

- Vary the duty cycle from 50 to 100 percent, which is set by the potentiometer i.e., 50% duty cycle – 0 VDC, 90% duty cycle is equal to 5 VDC.

- Display the motor RPM and duty cycle on an AND671GST LCD.

- Load the motor and perform compensation to return the RPMs to original value.

8.4.2   STRUCTURE CHART

The structure chart for the motor speed control project is shown in Figure 8.6.

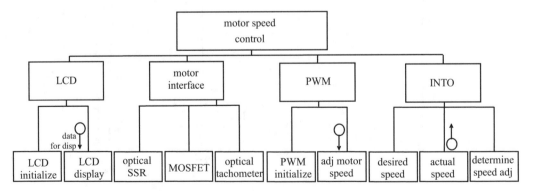

Figure 8.6: Structure chart for the motor speed control project.

8.4.3   UML ACTIVITY DIAGRAMS

The UML activity diagrams for the motor speed control project are shown in Figure 8.7.

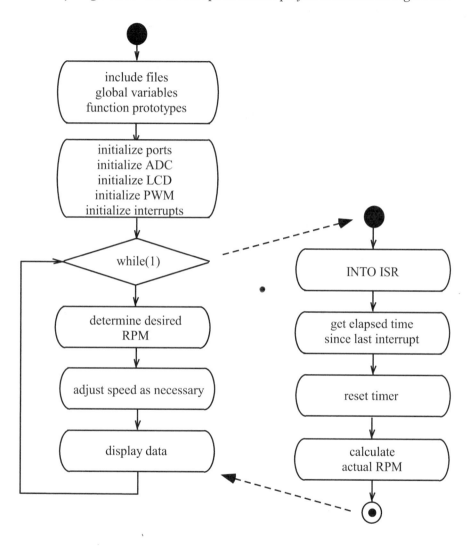

Figure 8.7: UML activity diagrams for the motor speed control project.

8.4.4   MICROCONTROLLER CODE

The code for the motor speed control project follows. Note that this code was implemented using the gcc AVR compiler. Note the different notation used for including the header file and configuring interrupts.

```
//**************************************************************************
//Code written by Geoff Luke, MSEE
//Last Updated: September 15, 2009
//**************************************************************************
//
//Note: This program was written using the gcc AVR compiler
//
//Description: This program powers a motor to run at 1125 to 2025 rpm. The
//value is set by a potentiometer.The microcontroller receives feedback from
//an optical tachometer that triggers external interrupt 0.  The desired
//speed is maintained even when the motor is loaded.
//
//Port connection:
//  Port C 0-7: used as data output to LCD
//  Port D 6-7: control pins for LCD
//  Port D 2:   External interrupt 0 (from tachometer)
//  Port A 0:   Used as ATD input
//  Port B 0:   PWM output            •
//
//**************************************************************************

//include files*************************************************************
#include <avr\io.h>
#include <avr\interrupt.h>

//function prototypes*******************************************************
void initialize_ports();
void initialize_ADC();
unsigned int readADC(unsigned char);
void LCD_init();
void putChar(unsigned char);
void putcommand(unsigned char);
void PWM_init(void);
void display_data(void);
void delay_5ms();

//global variables**********************************************************
unsigned int actualRPM, desiredRPM;
```

```
ISR(INT0_vect)
{
unsigned int time = TCNT0;
TCNT0 = 0x00;
actualRPM = 3906/time*60;
}

//*************************************************************************
int main(void)
{
initializePorts();
initializeADC();
LCD_init();
PWM_init();

MCUCR = 0x02;
GICR = 0x40;
TCCR0 = 0x05;
actualRPM = 0;
sei();

while(1)
  {
  desiredRPM = (unsigned int)(0.878906*(float)readADC(0)+1125.0);
  if(desiredRPM > actualRPM && OCR1BL != 0xFF)
    {
    OCR1BL = OCR1BL+1;
    }
  else if(desiredRPM < actualRPM && OCR1BL != 0x00)
    {
    OCR1BL = OCR1BL-1;
    }
  else
    {
    OCR1BL = OCR1BL;
    }
  displayData();
  }
return 0;
```

```
}

//************************************************************************

void initialize_ports()
{
DDRD = 0xFB;
DDRC = 0xFF;
DDRB = 0xFF;
}

//************************************************************************

void initialize_ports()
{
DDRD = 0xFB;
DDRC = 0xFF;
DDRB = 0xFF;

}

//************************************************************************


void initialize_ADC()
{
ADMUX = 0;                              //Select channel 0

 //Enable ADC and set module enable ADC
ADCSRA = 0xC3;                          //Set module prescalar to 8
while(!(ADCSRA & 0x10));                //Wait until conversion is ready
ADCSRA |= 0x10;                         //Clear conversion ready flag
}

//************************************************************************


unsigned int readADC(unsigned char channel)
{
```

```
unsigned int binary_weighted_voltage, binary_weighted_voltage_low;
unsigned int binary_weighted_voltage_high;//weighted binary voltage

ADMUX = channel;                        //Select channel
ADCSRA |= 0x43;                         //Start conversion
                                        //Set ADC module prescalar
                                        //to 8 critical for
                                        //accurate ADC results
while (!(ADCSRA & 0x10));                //Check if conversion is ready
ADCSRA |= 0x10;
 //Clear conv rdy flag - set the bit
binary_weighted_voltage_low = ADCL;
 //Read 8 low bits first (important)

 //Read 2 high bits, multiply by 256
binary_weighted_voltage_high = ((unsigned int)(ADCH << 8));
binary_weighted_voltage = binary_weighted_voltage_low
+ binary_weighted_voltage_high;
return binary_weighted_voltage;         //ADCH:ADCL
}


//*****************************************************************************
//LCD_Init: initialization for an LCD connected in the following manner:
//LCD: AND671GST 1x16 character display
//LCD configured as two 8 character lines in a 1x16 array
//LCD data bus (pin 14-pin7) ATMEL ATmega16: PORTC
//LCD RS (pin~(4) ATMEL ATmega16: PORTD[7]
//LCD E (pin~(6) ATMEL ATmega16: PORTD[6]
//*****************************************************************************

void LCD_init(void)


{
delay(1);
delay(1);
delay(1);
                                        //Output command string to
                                        //Initialize LCD
```

```
putcommand(0x38);                    //Function set 8-bit
delay(1);
putcommand(0x38);                    //Function set 8-bit
delay(1);
putcommand(0x38);                    //Function set 8-bit
putcommand(0x38);                    //One line, 5x7 char
putcommand(0x0E);                    //Display on
putcommand(0x01);                    //Display clear-1.64 ms
putcommand(0x06);                    //Entry mode set
putcommand(0x00);                    //Clear display, cursor at home
putcommand(0x00);                    //Clear display, cursor at home
}


//************************************************************************


void putChar(unsigned char~(c)
{
DDRC = 0xff;                         //Set PORTC as output
DDRD = DDRD|0xC0;                    //Make PORTD[7:6] output
PORTC = c;
PORTD = PORTD|0x80;                  //RS=1
PORTD = PORTD|0x40;                  //E=1
PORTD = PORTD&0xbf;                  //E=0
delay(1);
}



//************************************************************************


void putcommand(unsigned char~(d)
{


DDRC = 0xff;                         //Set PORTC as output
DDRD = DDRD|0xC0;                    //Make PORTD[7:6] output
PORTD = PORTD&0x7f;                  //RS=0
PORTC = d;
PORTD = PORTD|0x40; //E=1
PORTD = PORTD&0xbf; //E=0
```

```
delay(1);
}


//*********************************************************************
void display_data(void)
{
unsigned int thousands, hundreds, tens, ones, dutyCycle;

thousands = desiredRPM/1000;
hundreds = (desiredRPM - 1000*thousands)/100;
tens = (desiredRPM - 1000*thousands - 100*hundreds)/10;
ones = (desiredRPM - 1000*thousands - 100*hundreds - 10*tens);

putcommand(0x80);
putChar((unsigned char)(thousands)+48);
putChar((unsigned char)(hundreds)+48);
putChar((unsigned char)(tens)+48);

putChar((unsigned char)(ones)+48);
putChar('R');
putChar('P');
putChar('M');
putcommand(0xC0);

thousands = actualRPM/1000;
hundreds = (actualRPM - 1000*thousands)/100;
tens = (actualRPM - 1000*thousands - 100*hundreds)/10;
ones = (actualRPM - 1000*thousands - 100*hundreds - 10*tens);

putcommand(0xC0);
putChar((unsigned char)(thousands)+48);
putChar((unsigned char)(hundreds)+48);
putChar((unsigned char)(tens)+48);
putChar((unsigned char)(ones)+48);
putChar('R');
putChar('P');
putChar('M');
```

```
putChar(' ');
putChar(' ');

dutyCycle = OCR1BL*100/255;

hundreds = (dutyCycle)/100;
tens = (dutyCycle - 100*hundreds)/10;
ones = (dutyCycle - 100*hundreds - 10*tens);

if(hundreds > 0)
  {
  putChar((unsigned char)(hundreds)+48);
  }

else

  {
  putChar(' ');
  }
putChar((unsigned char)(tens)+48);
putChar((unsigned char)(ones)+48);
putChar('%');
}


//************************************************************************
void PWM_init(void)
{
unsigned int Open_Speed_int;
float Open_Speed_float;
int PWM_duty_cycle;


Open_Speed_int = readADC(0x02);          //Open Speed Setting
                                         //unsigned int
                                         //Convert to max duty
                                         //Cycle setting 0 VDC =
                                         //50% = 127, 5 VDC =
                                         //100% = 255
```

```
Open_Speed_float = ((float)(Open_Speed_int)/(float)(0x0400));
                                         //Convert volt to
                                         //PWM constant 127-255
Open_Speed_int = (unsigned int)((Open_Speed_float * 127) + 128.0);
                                         //Configure PWM clock
TCCR1A = 0xA1;                           //freq = resonator/510
                                         // = 4 MHz/510
                                         //freq = 19.607 kHz
TCCR1B = 0x02;                           //Clock source
                                         //division of 8
                                         //Initiate PWM duty cycle
                                         //variables
PWM_duty_cycle = 255;
OCR1BH = 0x00;
OCR1BL = (unsigned char)(PWM_duty_cycle); //set PWM duty cycle CH
}


//**************************************************************************
```

8.5   AUTONOMOUS MAZE NAVIGATING ROBOT

8.5.1   DESCRIPTION

Graymark (www.graymarkint.com) manufactures many low-cost, excellent robot platforms. In this project, we modify the Blinky 602A robot to be controlled by an ATmega164. The Blinky 602A kit contains the hardware and mechanical parts to construct a line following robot. The processing electronics for the robot consists of analog circuitry. The robot is controlled by two 3 VDC motors which independently drive a left wheel and a right wheel. A third non-powered drag wheel provides tripod stability for the robot.

In this project, we equip the Blinky 602A robot platform with three Sharp GP12D IR sensors as shown in Figure 8.8. The characteristics of the sensor are also shown. The robot is placed in a maze with reflective walls. The goal of the project is for the robot to detect wall placement and navigate through the maze. The robot does not have apriori knowledge about the configuration of the maze. The control algorithm for the robot is hosted on the ATmega164.

8.5.2   REQUIREMENTS

The requirements for this project are simple: the robot must autonomously navigate through the maze without touching maze walls.

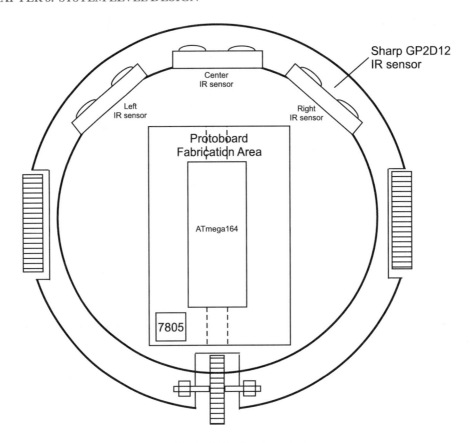

a) Graymark Blinky 602 robot layout - topview

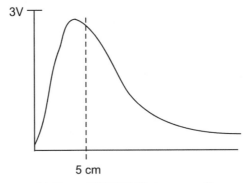

b) Sharp GP2D12 IR sensor profile

Figure 8.8: Robot layout.

8.5.3   CIRCUIT DIAGRAM

The circuit diagram for the robot is shown in Figure 8.9. The three IR sensors (left, middle, and right) will be mounted on the leading edge of the robot to detect maze walls. The output from the sensors are fed to three ADC channels (PORTA[2:0]). The robot motors are driven by PWM signals generated at the PWM channels A and B (OC1A and OC1B). The microcontroller is interfaced to the motors via a transistor with enough drive capability to handle the maximum current requirements of the motor. Since the microcontroller is powered at 5 VDC and the motors are rated at 3 VDC, two 1N4001 diodes are placed in series with the motor. This reduces the supply voltage to the motor to be approximately 3 VDC. The robot is powered by a 9 VDC battery which is fed to a 5 VDC voltage regulator.

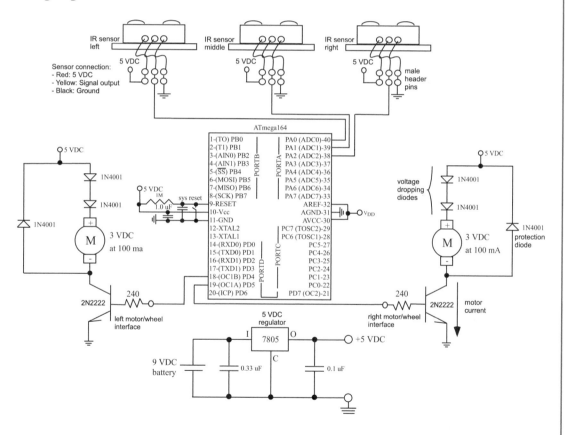

Figure 8.9: Robot circuit diagram.

8.5.4   STRUCTURE CHART

The structure chart for the robot project is provided in Figure 8.10.

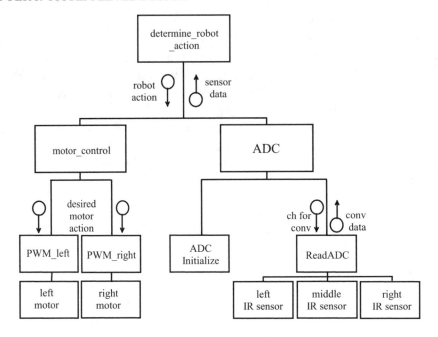

Figure 8.10: Robot structure diagram.

8.5.5   UML ACTIVITY DIAGRAMS
The UML activity diagram for the robot is provided in Figure 8.11.

8.5.6   MICROCONTROLLER CODE
We leave the microcontroller code as a homework assignment for the reader. We have provided all of the required functions throughout the text except for the function to link the IR sensor values to the robot appropriate robot decisions.

8.6   CHAPTER PROBLEMS

8.1.   Construct the UML activity diagrams for all functions related to the weather station.

8.2.   It is desired to updated weather parameters every 15 minutes. Write a function to provide a 15 minute delay.

8.3.   Add one of the following sensors to the weather station:

- anemometer
- barometer
- hygrometer

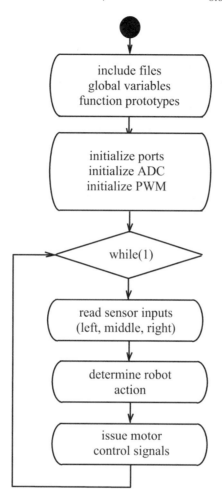

Figure 8.11: Robot UML activity diagram.

- rain gauge

- thermocouple

You will need to investigate background information on the selected sensor, develop an interface circuit for the sensor, and modify the weather station code.

8.4. Modify the motor speed control interface circuit to provide for bi-directional motor control.

8.5. Use optical encoder output channels A and B to determine motor direction and speed.

8.6. Modify the motor speed control algorithm to display motor direction (CW or CCW) and speed in RPM on the LCD.

8.7. The Blinky 602A robot under ATmega164 control abruptly starts and stops when PWM is applied. Modify the algorithm for a gradual ramp up (and down) of the motor speed.

8.8. Modify the Blinky 602A circuit and microcontroller code such that the maximum speed of the robot is set with an external potentiometer.

8.9. Modify the Blinky 602A circuit and microcontroller code such that the IR sensors are only asserted just before a range reading is taken.

8.10. Apply embedded system design techniques presented throughout the text to a project of your choosing. Follow the design process and provide the following products:

- system description,
- system requirements,
- a structure chart,
- system circuit diagram,
- UML activity diagrams, and the
- microcontroller code.

8.11. Add the following features to the Blinky 602A platform:

- Line following capability (Hint: adapt the line following circuitry onboard the Blinky 602A to operate with the ATmega164.)
- Two way robot communications (use the onboard IR sensors)
- LCD display for status and troubleshooting display
- Voice output (Hint: use an ISD 4003 Chip Corder)

8.12. Develop an embedded system controlled submarine (`www.seaperch.org`).

8.13. Equip the ATmega164 with automatic cell phone dialing capability to notify you when a fire is present in your home.

8.14. Develop an embedded system controlled dirigible/blimp (`www.microflight.com`, `www.rctoys.com`).

8.15. Develop a trip odometer for your bicycle (Hint: use a Hall Effect sensor to detect tire rotation).

8.16. Develop a timing system for a 4 lane pinewood derby track.

APPENDIX A

ATmega164 Register Set

Address	Name	Bit 7	Bit 6	Bit 5	Bit 4	Bit 3	Bit 2	Bit 1	Bit 0	Page
(0xFF)	Reserved	-	-	-	-	-	-	-	-	
(0xFE)	Reserved	-	-	-	-	-	-	-	-	
(0xFD)	Reserved	-	-	-	-	-	-	-	-	
(0xFC)	Reserved	-	-	-	-	-	-	-	-	
(0xFB)	Reserved	-	-	-	-	-	-	-	-	
(0xFA)	Reserved	-	-	-	-	-	-	-	-	
(0xF9)	Reserved	-	-	-	-	-	-	-	-	
(0xF8)	Reserved	-	-	-	-	-	-	-	-	
(0xF7)	Reserved	-	-	-	-	-	-	-	-	
(0xF6)	Reserved	-	-	-	-	-	-	-	-	
(0xF5)	Reserved	-	-	-	-	-	-	-	-	
(0xF4)	Reserved	-	-	-	-	-	-	-	-	
(0xF3)	Reserved	-	-	-	-	-	-	-	-	
(0xF2)	Reserved	-	-	-	-	-	-	-	-	
(0xF1)	Reserved	-	-	-	-	-	-	-	-	
(0xF0)	Reserved	-	-	-	-	-	-	-	-	
(0xEF)	Reserved	-	-	-	-	-	-	-	-	
(0xEE)	Reserved	-	-	-	-	-	-	-	-	
(0xED)	Reserved	-	-	-	-	-	-	-	-	
(0xEC)	Reserved	-	-	-	-	-	-	-	-	
(0xEB)	Reserved	-	-	-	-	-	-	-	-	
(0xEA)	Reserved	-	-	-	-	-	-	-	-	
(0xE9)	Reserved	-	-	-	-	-	-	-	-	
(0xE8)	Reserved	-	-	-	-	-	-	-	-	
(0xE7)	Reserved	-	-	-	-	-	-	-	-	
(0xE6)	Reserved	-	-	-	-	-	-	-	-	
(0xE5)	Reserved	-	-	-	-	-	-	-	-	
(0xE4)	Reserved	-	-	-	-	-	-	-	-	
(0xE3)	Reserved	-	-	-	-	-	-	-	-	
(0xE2)	Reserved	-	-	-	-	-	-	-	-	
(0xE1)	Reserved	-	-	-	-	-	-	-	-	
(0xE0)	Reserved	-	-	-	-	-	-	-	-	
(0xDF)	Reserved	-	-	-	-	-	-	-	-	
(0xDE)	Reserved	-	-	-	-	-	-	-	-	
(0xDD)	Reserved	-	-	-	-	-	-	-	-	
(0xDC)	Reserved	-	-	-	-	-	-	-	-	
(0xDB)	Reserved	-	-	-	-	-	-	-	-	
(0xDA)	Reserved	-	-	-	-	-	-	-	-	
(0xD9)	Reserved	-	-	-	-	-	-	-	-	
(0xD8)	Reserved	-	-	-	-	-	-	-	-	
(0xD7)	Reserved	-	-	-	-	-	-	-	-	
(0xD6)	Reserved	-	-	-	-	-	-	-	-	
(0xD5)	Reserved	-	-	-	-	-	-	-	-	
(0xD4)	Reserved	-	-	-	-	-	-	-	-	
(0xD3)	Reserved	-	-	-	-	-	-	-	-	
(0xD2)	Reserved	-	-	-	-	-	-	-	-	
(0xD1)	Reserved	-	-	-	-	-	-	-	-	
(0xD0)	Reserved	-	-	-	-	-	-	-	-	
(0xCF)	Reserved	-	-	-	-	-	-	-	-	
(0xCE)	UDR1	USART1 I/O Data Register								190
(0xCD)	UBRR1H	-	-	-	-	USART1 Baud Rate Register High Byte				194/207
(0xCC)	UBRR1L	USART1 Baud Rate Register Low Byte								194/207
(0xCB)	Reserved	-	-	-	-	-	-	-	-	
(0xCA)	UCSR1C	UMSEL11	UMSEL10	-	-	-	UDORD1	UCPHA1	UCPOL1	192/206
(0xC9)	UCSR1B	RXCIE1	TXCIE1	UDRIE1	RXEN1	TXEN1	UCSZ12	RXB81	TXB81	191/205
(0xC8)	UCSR1A	RXC1	TXC1	UDRE1	FE1	DOR1	UPE1	U2X1	MPCM1	190/205
(0xC7)	Reserved	-	-	-	-	-	-	-	-	
(0xC6)	UDR0	USART0 I/O Data Register								190
(0xC5)	UBRR0H	-	-	-	-	USART0 Baud Rate Register High Byte				194/207
(0xC4)	UBRR0L	USART0 Baud Rate Register Low Byte								194/207
(0xC3)	Reserved	-	-	-	-	-	-	-	-	
(0xC2)	UCSR0C	UMSEL01	UMSEL00	-	-	-	UDORD0	UCPHA0	UCPOL0	192/206
(0xC1)	UCSR0B	RXCIE0	TXCIE0	UDRIE0	RXEN0	TXEN0	UCSZ02	RXB80	TXB80	191/205

Figure A.1: Atmel AVR ATmega164 Register Set. (Figure used with permission of Atmel, Incorporated.)

Address	Name	Bit 7	Bit 6	Bit 5	Bit 4	Bit 3	Bit 2	Bit 1	Bit 0	Page
(0xC0)	UCSR0A	RXC0	TXC0	UDRE0	FE0	DOR0	UPE0	U2X0	MPCM0	190/205
(0xBF)	Reserved	-	-	-	-	-	-	-	-	
(0xBE)	Reserved	-	-	-	-	-	-	-	-	
(0xBD)	TWAMR	TWAM6	TWAM5	TWAM4	TWAM3	TWAM2	TWAM1	TWAM0	-	236
(0xBC)	TWCR	TWINT	TWEA	TWSTA	TWSTO	TWWC	TWEN	-	TWIE	233
(0xBB)	TWDR	2-wire Serial Interface Data Register								235
(0xBA)	TWAR	TWA6	TWA5	TWA4	TWA3	TWA2	TWA1	TWA0	TWGCE	236
(0xB9)	TWSR	TWS7	TWS6	TWS5	TWS4	TWS3	-	TWPS1	TWPS0	235
(0xB8)	TWBR	2-wire Serial Interface Bit Rate Register								233
(0xB7)	Reserved	-	-	-	-	-	-	-	-	
(0xB6)	ASSR	-	EXCLK	AS2	TCN2UB	OCR2AUB	OCR2BUB	TCR2AUB	TCR2BUB	158
(0xB5)	Reserved	-	-	-	-	-	-	-	-	
(0xB4)	OCR2B	Timer/Counter2 Output Compare Register B								158
(0xB3)	OCR2A	Timer/Counter2 Output Compare Register A								158
(0xB2)	TCNT2	Timer/Counter2 (8 Bit)								157
(0xB1)	TCCR2B	FOC2A	FOC2B	-	-	WGM22	CS22	CS21	CS20	156
(0xB0)	TCCR2A	COM2A1	COM2A0	COM2B1	COM2B0	-	-	WGM21	WGM20	153
(0xAF)	Reserved	-	-	-	-	-	-	-	-	
(0xAE)	Reserved	-	-	-	-	-	-	-	-	
(0xAD)	Reserved	-	-	-	-	-	-	-	-	
(0xAC)	Reserved	-	-	-	-	-	-	-	-	
(0xAB)	Reserved	-	-	-	-	-	-	-	-	
(0xAA)	Reserved	-	-	-	-	-	-	-	-	
(0xA9)	Reserved	-	-	-	-	-	-	-	-	
(0xA8)	Reserved	-	-	-	-	-	-	-	-	
(0xA7)	Reserved	-	-	-	-	-	-	-	-	
(0xA6)	Reserved	-	-	-	-	-	-	-	-	
(0xA5)	Reserved	-	-	-	-	-	-	-	-	
(0xA4)	Reserved	-	-	-	-	-	-	-	-	
(0xA3)	Reserved	-	-	-	-	-	-	-	-	
(0xA2)	Reserved	-	-	-	-	-	-	-	-	
(0xA1)	Reserved	-	-	-	-	-	-	-	-	
(0xA0)	Reserved	-	-	-	-	-	-	-	-	
(0x9F)	Reserved	-	-	-	-	-	-	-	-	
(0x9E)	Reserved	-	-	-	-	-	-	-	-	
(0x9D)	Reserved	-	-	-	-	-	-	-	-	
(0x9C)	Reserved	-	-	-	-	-	-	-	-	
(0x9B)	Reserved	-	-	-	-	-	-	-	-	
(0x9A)	Reserved	-	-	-	-	-	-	-	-	
(0x99)	Reserved	-	-	-	-	-	-	-	-	
(0x98)	Reserved	-	-	-	-	-	-	-	-	
(0x97)	Reserved	-	-	-	-	-	-	-	-	
(0x96)	Reserved	-	-	-	-	-	-	-	-	
(0x95)	Reserved	-	-	-	-	-	-	-	-	
(0x94)	Reserved	-	-	-	-	-	-	-	-	
(0x93)	Reserved	-	-	-	-	-	-	-	-	
(0x92)	Reserved	-	-	-	-	-	-	-	-	
(0x91)	Reserved	-	-	-	-	-	-	-	-	
(0x90)	Reserved	-	-	-	-	-	-	-	-	
(0x8F)	Reserved	-	-	-	-	-	-	-	-	
(0x8E)	Reserved	-	-	-	-	-	-	-	-	
(0x8D)	Reserved	-	-	-	-	-	-	-	-	
(0x8C)	Reserved	-	-	-	-	-	-	-	-	
(0x8B)	OCR1BH	Timer/Counter1 - Output Compare Register B High Byte								137
(0x8A)	OCR1BL	Timer/Counter1 - Output Compare Register B Low Byte								137
(0x89)	OCR1AH	Timer/Counter1 - Output Compare Register A High Byte								137
(0x88)	OCR1AL	Timer/Counter1 - Output Compare Register A Low Byte								137
(0x87)	ICR1H	Timer/Counter1 - Input Capture Register High Byte								138
(0x86)	ICR1L	Timer/Counter1 - Input Capture Register Low Byte								138
(0x85)	TCNT1H	Timer/Counter1 - Counter Register High Byte								137
(0x84)	TCNT1L	Timer/Counter1 - Counter Register Low Byte								137
(0x83)	Reserved	-	-	-	-	-	-	-	-	
(0x82)	TCCR1C	FOC1A	FOC1B	-	-	-	-	-	-	136
(0x81)	TCCR1B	ICNC1	ICES1	-	WGM13	WGM12	CS12	CS11	CS10	135
(0x80)	TCCR1A	COM1A1	COM1A0	COM1B1	COM1B0	-	-	WGM11	WGM10	133
(0x7F)	DIDR1	-	-	-	-	-	-	AIN1D	AIN0D	240

Figure A.2: Atmel AVR ATmega164 Register Set. (Figure used with permission of Atmel, Incorporated.)

Address	Name	Bit 7	Bit 6	Bit 5	Bit 4	Bit 3	Bit 2	Bit 1	Bit 0	Page
(0x7E)	DIDR0	ADC7D	ADC6D	ADC5D	ADC4D	ADC3D	ADC2D	ADC1D	ADC0D	260
(0x7D)	Reserved	-	-	-	-	-	-	-	-	
(0x7C)	ADMUX	REFS1	REFS0	ADLAR	MUX4	MUX3	MUX2	MUX1	MUX0	256
(0x7B)	ADCSRB	-	ACME	-	-	-	ADTS2	ADTS1	ADTS0	239
(0x7A)	ADCSRA	ADEN	ADSC	ADATE	ADIF	ADIE	ADPS2	ADPS1	ADPS0	258
(0x79)	ADCH	ADC Data Register High byte								259
(0x78)	ADCL	ADC Data Register Low byte								259
(0x77)	Reserved	-	-	-	-	-	-	-	-	
(0x76)	Reserved	-	-	-	-	-	-	-	-	
(0x75)	Reserved	-	-	-	-	-	-	-	-	
(0x74)	Reserved	-	-	-	-	-	-	-	-	
(0x73)	PCMSK3	PCINT31	PCINT30	PCINT29	PCINT28	PCINT27	PCINT26	PCINT25	PCINT24	71
(0x72)	Reserved	-	-	-	-	-	-	-	-	
(0x71)	Reserved	-	-	-	-	-	-	-	-	
(0x70)	TIMSK2	-	-	-	-	-	OCIE2B	OCIE2A	TOIE2	159
(0x6F)	TIMSK1	-	-	ICIE1	-	-	OCIE1B	OCIE1A	TOIE1	138
(0x6E)	TIMSK0	-	-	-	-	-	OCIE0B	OCIE0A	TOIE0	110
(0x6D)	PCMSK2	PCINT23	PCINT22	PCINT21	PCINT20	PCINT19	PCINT18	PCINT17	PCINT16	71
(0x6C)	PCMSK1	PCINT15	PCINT14	PCINT13	PCINT12	PCINT11	PCINT10	PCINT9	PCINT8	71
(0x6B)	PCMSK0	PCINT7	PCINT6	PCINT5	PCINT4	PCINT3	PCINT2	PCINT1	PCINT0	72
(0x6A)	Reserved	-	-	-	-	-	-	-	-	
(0x69)	EICRA	-	-	ISC21	ISC20	ISC11	ISC10	ISC01	ISC00	68
(0x68)	PCICR	-	-	-	-	PCIE3	PCIE2	PCIE1	PCIE0	70
(0x67)	Reserved	-	-	-	-	-	-	-	-	
(0x66)	OSCCAL	Oscillator Calibration Register								41
(0x65)	Reserved	-	-	-	-	-	-	-	-	
(0x64)	PRR	PRTWI	PRTIM2	PRTIM0	PRUSART1	PRTIM1	PRSPI	PRUSART0	PRADC	49
(0x63)	Reserved	-	-	-	-	-	-	-	-	
(0x62)	Reserved	-	-	-	-	-	-	-	-	
(0x61)	CLKPR	CLKPCE	-	-	-	CLKPS3	CLKPS2	CLKPS1	CLKPS0	41
(0x60)	WDTCSR	WDIF	WDIE	WDP3	WDCE	WDE	WDP2	WDP1	WDP0	60
0x3F (0x5F)	SREG	I	T	H	S	V	N	Z	C	11
0x3E (0x5E)	SPH	SP15	SP14	SP13	SP12	SP11	SP10	SP9	SP8	12
0x3D (0x5D)	SPL	SP7	SP6	SP5	SP4	SP3	SP2	SP1	SP0	12
0x3C (0x5C)	Reserved	-	-	-	-	-	-	-	-	
0x3B (0x5B)	RAMPZ	-	-	-	-	-	-	-	RAMPZ0	15
0x3A (0x5A)	Reserved	-	-	-	-	-	-	-	-	
0x39 (0x59)	Reserved	-	-	-	-	-	-	-	-	
0x38 (0x58)	Reserved	-	-	-	-	-	-	-	-	
0x37 (0x57)	SPMCSR	SPMIE	RWWSB	SIGRD	RWWSRE	BLBSET	PGWRT	PGERS	SPMEN	202
0x36 (0x56)	Reserved	-	-	-	-	-	-	-	-	
0x35 (0x55)	MCUCR	JTD	BODS	BODSE	PUD	-	-	IVSEL	IVCE	92/276
0x34 (0x54)	MCUSR	-	-	-	JTRF	WDRF	BORF	EXTRF	PORF	59/276
0x33 (0x53)	SMCR	-	-	-	-	SM2	SM1	SM0	SE	48
0x32 (0x52)	Reserved	-	-	-	-	-	-	-	-	
0x31 (0x51)	OCDR	On-Chip Debug Register								266
0x30 (0x50)	ACSR	ACD	ACBG	ACO	ACI	ACIE	ACIC	ACIS1	ACIS0	258
0x2F (0x4F)	Reserved	-	-	-	-	-	-	-	-	
0x2E (0x4E)	SPDR	SPI 0 Data Register								171
0x2D (0x4D)	SPSR	SPIF0	WCOL0	-	-	-	-	-	SPI2X0	170
0x2C (0x4C)	SPCR	SPIE0	SPE0	DORD0	MSTR0	CPOL0	CPHA0	SPR01	SPR00	169
0x2B (0x4B)	GPIOR2	General Purpose I/O Register 2								29
0x2A (0x4A)	GPIOR1	General Purpose I/O Register 1								29
0x29 (0x49)	Reserved	-	-	-	-	-	-	-	-	
0x28 (0x48)	OCR0B	Timer/Counter0 Output Compare Register B								110
0x27 (0x47)	OCR0A	Timer/Counter0 Output Compare Register A								109
0x26 (0x46)	TCNT0	Timer/Counter0 (8 Bit)								109
0x25 (0x45)	TCCR0B	FOC0A	FOC0B	-	-	WGM02	CS02	CS01	CS00	108
0x24 (0x44)	TCCR0A	COM0A1	COM0A0	COM0B1	COM0B0	-	-	WGM01	WGM00	110
0x23 (0x43)	GTCCR	TSM	-	-	-	-	-	PSRASY	PSRSYNC	160
0x22 (0x42)	EEARH	-	-	-	-	EEPROM Address Register High Byte				24
0x21 (0x41)	EEARL	EEPROM Address Register Low Byte								24
0x20 (0x40)	EEDR	EEPROM Data Register								24
0x1F (0x3F)	EECR	-	-	EEPM1	EEPM0	EERIE	EEMPE	EEPE	EERE	24
0x1E (0x3E)	GPIOR0	General Purpose I/O Register 0								29
0x1D (0x3D)	EIMSK	-	-	-	-	-	INT2	INT1	INT0	69

Figure A.3: Atmel AVR ATmega164 Register Set. (Figure used with permission of Atmel, Incorporated.)

Address	Name	Bit 7	Bit 6	Bit 5	Bit 4	Bit 3	Bit 2	Bit 1	Bit 0	Page
0x1C (0x3C)	EIFR	-	-	-	-	-	INTF2	INTF1	INTF0	69
0x1B (0x3B)	PCIFR	-	-	-	-	PCIF3	PCIF2	PCIF1	PCIF0	70
0x1A (0x3A)	Reserved	-	-	-	-	-	-	-	-	
0x19 (0x39)	Reserved	-	-	-	-	-	-	-	-	
0x18 (0x38)	Reserved	-	-	-	-	-	-	-	-	
0x17 (0x37)	TIFR2	-	-	-	-	-	OCF2B	OCF2A	TOV2	160
0x16 (0x36)	TIFR1	-	-	ICF1	-	-	OCF1B	OCF1A	TOV1	139
0x15 (0x35)	TIFR0	-	-	-	-	-	OCF0B	OCF0A	TOV0	110
0x14 (0x34)	Reserved	-	-	-	-	-	-	-	-	
0x13 (0x33)	Reserved	-	-	-	-	-	-	-	-	
0x12 (0x32)	Reserved	-	-	-	-	-	-	-	-	
0x11 (0x31)	Reserved	-	-	-	-	-	-	-	-	
0x10 (0x30)	Reserved	-	-	-	-	-	-	-	-	
0x0F (0x2F)	Reserved	-	-	-	-	-	-	-	-	
0x0E (0x2E)	Reserved	-	-	-	-	-	-	-	-	
0x0D (0x2D)	Reserved	-	-	-	-	-	-	-	-	
0x0C (0x2C)	Reserved	-	-	-	-	-	-	-	-	
0x0B (0x2B)	PORTD	PORTD7	PORTD6	PORTD5	PORTD4	PORTD3	PORTD2	PORTD1	PORTD0	93
0x0A (0x2A)	DDRD	DDD7	DDD6	DDD5	DDD4	DDD3	DDD2	DDD1	DDD0	93
0x09 (0x29)	PIND	PIND7	PIND6	PIND5	PIND4	PIND3	PIND2	PIND1	PIND0	93
0x08 (0x28)	PORTC	PORTC7	PORTC6	PORTC5	PORTC4	PORTC3	PORTC2	PORTC1	PORTC0	93
0x07 (0x27)	DDRC	DDC7	DDC6	DDC5	DDC4	DDC3	DDC2	DDC1	DDC0	93
0x06 (0x26)	PINC	PINC7	PINC6	PINC5	PINC4	PINC3	PINC2	PINC1	PINC0	93
0x05 (0x25)	PORTB	PORTB7	PORTB6	PORTB5	PORTB4	PORTB3	PORTB2	PORTB1	PORTB0	92
0x04 (0x24)	DDRB	DDB7	DDB6	DDB5	DDB4	DDB3	DDB2	DDB1	DDB0	92
0x03 (0x23)	PINB	PINB7	PINB6	PINB5	PINB4	PINB3	PINB2	PINB1	PINB0	92
0x02 (0x22)	PORTA	PORTA7	PORTA6	PORTA5	PORTA4	PORTA3	PORTA2	PORTA1	PORTA0	92
0x01 (0x21)	DDRA	DDA7	DDA6	DDA5	DDA4	DDA3	DDA2	DDA1	DDA0	92
0x00 (0x20)	PINA	PINA7	PINA6	PINA5	PINA4	PINA3	PINA2	PINA1	PINA0	92

Notes:
1. For compatibility with future devices, reserved bits should be written to zero if accessed. Reserved I/O memory addresses should never be written.
2. I/O registers within the address range $00 - $1F are directly bit-accessible using the SBI and CBI instructions. In these registers, the value of single bits can be checked by using the SBIS and SBIC instructions.
3. Some of the status flags are cleared by writing a logical one to them. Note that the CBI and SBI instructions will operate on all bits in the I/O register, writing a one back into any flag read as set, thus clearing the flag. The CBI and SBI instructions work with registers 0x00 to 0x1F only.
4. When using the I/O specific commands IN and OUT, the I/O addresses $00 - $3F must be used. When addressing I/O registers as data space using LD and ST instructions, $20 must be added to these addresses. The ATmega164P/324P/644P is a complex microcontroller with more peripheral units than can be supported within the 64 location reserved in Opcode for the IN and OUT instructions. For the Extended I/O space from $60 - $FF, only the ST/STS/STD and LD/LDS/LDD instructions can be used.

Figure A.4: Atmel AVR ATmega164 Register Set. (Figure used with permission of Atmel, Incorporated.)

APPENDIX B

ATmega164 Header File

During C programming, the contents of a specific register may be referred to by name when an appropriate header file is included within your program. The header file provides the link between the register name used within a program and the hardware location of the register.

Provided below is the ATmega164 header file from the ICC AVR compiler. This header file was provided courtesy of ImageCraft Incorporated.

```
#ifndef __iom164pv_h
#define __iom164pv_h

/* ATmega164P header file for
 * ImageCraft ICCAVR compiler
 */

/* i/o register addresses
 * >= 0x60 are memory mapped only
 */

/* 2006/10/01 created
 */

/* Port D */
#define PIND (*(volatile unsigned char *)0x29)
#define DDRD (*(volatile unsigned char *)0x2A)
#define PORTD (*(volatile unsigned char *)0x2B)

/* Port C */
#define PINC (*(volatile unsigned char *)0x26)
#define DDRC (*(volatile unsigned char *)0x27)
#define PORTC (*(volatile unsigned char *)0x28)

/* Port B */
#define PINB (*(volatile unsigned char *)0x23)
#define DDRB (*(volatile unsigned char *)0x24)
#define PORTB (*(volatile unsigned char *)0x25)
```

```
/* Port A */
#define PINA (*(volatile unsigned char *)0x20)
#define DDRA (*(volatile unsigned char *)0x21)
#define PORTA (*(volatile unsigned char *)0x22)


/* Timer/Counter Interrupts */
#define TIFR0 (*(volatile unsigned char *)0x35)
#define  OCF0B    2
#define  OCF0A    1
#define  TOV0     0
#define TIMSK0 (*(volatile unsigned char *)0x6E)
#define  OCIE0B   2
#define  OCIE0A   1
#define  TOIE0    0
#define TIFR1 (*(volatile unsigned char *)0x36)
#define  ICF1     5
#define  OCF1B    2
#define  OCF1A    1
#define  TOV1     0
#define TIMSK1 (*(volatile unsigned char *)0x6F)
#define  ICIE1    5
#define  OCIE1B   2
#define  OCIE1A   1
#define  TOIE1    0
#define TIFR2 (*(volatile unsigned char *)0x37)
#define  OCF2B    2
#define  OCF2A    1
#define  TOV2     0
#define TIMSK2 (*(volatile unsigned char *)0x70)
#define  OCIE2B   2
#define  OCIE2A   1
#define  TOIE2    0


/* External Interrupts */
#define EIFR (*(volatile unsigned char *)0x3C)
#define  INTF2    2
#define  INTF1    1
#define  INTF0    0
```

```
#define EIMSK (*(volatile unsigned char *)0x3D)
#define   INT2     2
#define   INT1     1
#define   INT0     0
#define EICRA (*(volatile unsigned char *)0x69)
#define   ISC21    5
#define   ISC20    4
#define   ISC11    3
#define   ISC10    2
#define   ISC01    1
#define   ISC00    0


/* Pin Change Interrupts */
#define PCIFR (*(volatile unsigned char *)0x3B)
#define   PCIF3    3
#define   PCIF2    2
#define   PCIF1    1
#define   PCIF0    0
#define PCICR (*(volatile unsigned char *)0x68)
#define   PCIE3    3
#define   PCIE2    2
#define   PCIE1    1
#define   PCIE0    0
#define PCMSK0 (*(volatile unsigned char *)0x6B)
#define PCMSK1 (*(volatile unsigned char *)0x6C)
#define PCMSK2 (*(volatile unsigned char *)0x6D)
#define PCMSK3 (*(volatile unsigned char *)0x73)


/* GPIOR */
#define GPIOR0 (*(volatile unsigned char *)0x3E)
#define GPIOR1 (*(volatile unsigned char *)0x4A)
#define GPIOR2 (*(volatile unsigned char *)0x4B)

/* EEPROM */
#define EECR (*(volatile unsigned char *)0x3F)
#define   EEPM1    5
#define   EEPM0    4
#define   EERIE    3
#define   EEMPE    2
```

```
#define  EEMWE    2
#define  EEPE     1
#define  EEWE     1
#define  EERE     0
#define EEDR (*(volatile unsigned char *)0x40)
#define EEAR (*(volatile unsigned int *)0x41)
#define EEARL (*(volatile unsigned char *)0x41)
#define EEARH (*(volatile unsigned char *)0x42)

/* GTCCR */
#define GTCCR (*(volatile unsigned char *)0x43)
#define  TSM      7
#define  PSRASY   1
#define  PSR2     1
#define  PSRSYNC  0
#define  PSR10    0

/* Timer/Counter 0 */
#define OCR0B (*(volatile unsigned char *)0x48)
#define OCR0A (*(volatile unsigned char *)0x47)
#define TCNT0 (*(volatile unsigned char *)0x46)
#define TCCR0B (*(volatile unsigned char *)0x45)
#define  FOCOA    7
#define  FOCOB    6
#define  WGM02    3
#define  CS02     2
#define  CS01     1
#define  CS00     0
#define TCCR0A (*(volatile unsigned char *)0x44)
#define  COM0A1   7
#define  COM0A0   6
#define  COM0B1   5
#define  COM0B0   4
#define  WGM01    1
#define  WGM00    0

/* SPI */
#define SPCR (*(volatile unsigned char *)0x4C)
#define  SPIE     7
```

```
#define   SPE       6
#define   DORD      5
#define   MSTR      4
#define   CPOL      3
#define   CPHA      2
#define   SPR1      1
#define   SPR0      0
#define SPSR (*(volatile unsigned char *)0x4D)
#define   SPIF      7
#define   WCOL      6
#define   SPI2X     0
#define SPDR (*(volatile unsigned char *)0x4E)


/* Analog Comparator Control and Status Register */
#define ACSR (*(volatile unsigned char *)0x50)
#define   ACD       7
#define   ACBG      6
#define   ACO       5
#define   ACI       4
#define   ACIE      3
#define   ACIC      2
#define   ACIS1     1
#define   ACIS0     0


/* OCDR */
#define OCDR (*(volatile unsigned char *)0x51)
#define   IDRD      7


/* MCU */
#define MCUSR (*(volatile unsigned char *)0x54)
#define   JTRF      4
#define   WDRF      3
#define   BORF      2
#define   EXTRF     1
#define   PORF      0
#define MCUCR (*(volatile unsigned char *)0x55)
#define   JTD       7
#define   PUD       4
#define   IVSEL     1
```

```
#define   IVCE      0

#define SMCR (*(volatile unsigned char *)0x53)
#define   SM2       3
#define   SM1       2
#define   SM0       1
#define   SE        0

/* SPM Control and Status Register */
#define SPMCSR (*(volatile unsigned char *)0x57)
#define   SPMIE     7
#define   RWWSB     6
#define   SIGRD     5
#define   RWWSRE    4
#define   BLBSET    3
#define   PGWRT     2
#define   PGERS     1
#define   SPMEN     0


/* Stack Pointer */
#define SP   (*(volatile unsigned int *)0x5D)
#define SPL  (*(volatile unsigned char *)0x5D)
#define SPH  (*(volatile unsigned char *)0x5E)


/* Status REGister */
#define SREG (*(volatile unsigned char *)0x5F)


/* Watchdog Timer Control Register */
#define WDTCSR (*(volatile unsigned char *)0x60)
#define WDTCR (*(volatile unsigned char *)0x60)
#define   WDIF      7
#define   WDIE      6
#define   WDP3      5
#define   WDCE      4
#define   WDE       3
#define   WDP2      2
#define   WDP1      1
#define   WDP0      0
```

```
/* clock prescaler control register */
#define CLKPR (*(volatile unsigned char *)0x61)
#define   CLKPCE   7
#define   CLKPS3   3
#define   CLKPS2   2
#define   CLKPS1   1
#define   CLKPS0   0


/* PRR */
#define PRR0 (*(volatile unsigned char *)0x64)
#define   PRTWI    7
#define   PRTIM2   6
#define   PRTIM0   5
#define   PRUSART1 4
#define   PRTIM1   3
#define   PRSPI    2
#define   PRUSART0 1
#define   PRADC    0


/* Oscillator Calibration Register */
#define OSCCAL (*(volatile unsigned char *)0x66)

/* ADC */
#define ADC  (*(volatile unsigned int *)0x78)
#define ADCL (*(volatile unsigned char *)0x78)
#define ADCH (*(volatile unsigned char *)0x79)
#define ADCSRA (*(volatile unsigned char *)0x7A)
#define   ADEN     7
#define   ADSC     6
#define   ADATE    5
#define   ADIF     4
#define   ADIE     3
#define   ADPS2    2
#define   ADPS1    1
#define   ADPS0    0
#define ADCSRB (*(volatile unsigned char *)0x7B)
#define   ACME     6
#define   ADTS2    2
#define   ADTS1    1
```

```
#define  ADTS0     0
#define  ADMUX (*(volatile unsigned char *)0x7C)
#define  REFS1     7
#define  REFS0     6
#define  ADLAR     5
#define  MUX4      4
#define  MUX3      3
#define  MUX2      2
#define  MUX1      1
#define  MUX0      0

/* DIDR */
#define  DIDR0 (*(volatile unsigned char *)0x7E)
#define  ADC7D     7
#define  ADC6D     6
#define  ADC5D     5
#define  ADC4D     4
#define  ADC3D     3
#define  ADC2D     2
#define  ADC1D     1
#define  ADC0D     0
#define  DIDR1 (*(volatile unsigned char *)0x7F)
#define  AIN1D     1
#define  AIN0D     0

/* Timer/Counter1 */
#define  ICR1 (*(volatile unsigned int *)0x86)
#define  ICR1L (*(volatile unsigned char *)0x86)
#define  ICR1H (*(volatile unsigned char *)0x87)
#define  OCR1B (*(volatile unsigned int *)0x8A)
#define  OCR1BL (*(volatile unsigned char *)0x8A)
#define  OCR1BH (*(volatile unsigned char *)0x8B)
#define  OCR1A (*(volatile unsigned int *)0x88)
#define  OCR1AL (*(volatile unsigned char *)0x88)
#define  OCR1AH (*(volatile unsigned char *)0x89)
#define  TCNT1 (*(volatile unsigned int *)0x84)
#define  TCNT1L (*(volatile unsigned char *)0x84)
#define  TCNT1H (*(volatile unsigned char *)0x85)
#define  TCCR1C (*(volatile unsigned char *)0x82)
```

```
#define  FOC1A     7
#define  FOC1B     6
#define TCCR1B (*(volatile unsigned char *)0x81)
#define  ICNC1     7
#define  ICES1     6
#define  WGM13     4
#define  WGM12     3
#define  CS12      2
#define  CS11      1
#define  CS10      0
#define TCCR1A (*(volatile unsigned char *)0x80)
#define  COM1A1    7
#define  COM1A0    6
#define  COM1B1    5
#define  COM1B0    4
#define  WGM11     1
#define  WGM10     0

/* Timer/Counter2 */
#define ASSR (*(volatile unsigned char *)0xB6)
#define  EXCLK     6
#define  AS2       5
#define  TCN2UB    4
#define  OCR2AUB   3
#define  OCR2BUB   2
#define  TCR2AUB   1
#define  TCR2BUB   0
#define OCR2B (*(volatile unsigned char *)0xB4)
#define OCR2A (*(volatile unsigned char *)0xB3)
#define TCNT2 (*(volatile unsigned char *)0xB2)
#define TCCR2B (*(volatile unsigned char *)0xB1)
#define  FOC2A     7
#define  FOC2B     6
#define  WGM22     3
#define  CS22      2
#define  CS21      1
#define  CS20      0
#define TCCR2A (*(volatile unsigned char *)0xB0)
#define  COM2A1    7
```

```
#define  COM2A0    6
#define  COM2B1    5
#define  COM2B0    4
#define  WGM21     1
#define  WGM20     0


/* 2-wire SI */
#define TWBR (*(volatile unsigned char *)0xB8)
#define TWSR (*(volatile unsigned char *)0xB9)
#define  TWPS1     1
#define  TWPS0     0
#define TWAR (*(volatile unsigned char *)0xBA)
#define  TWGCE     0
#define TWDR (*(volatile unsigned char *)0xBB)
#define TWCR (*(volatile unsigned char *)0xBC)
#define  TWINT     7
#define  TWEA      6
#define  TWSTA     5
#define  TWSTO     4
#define  TWWC      3
#define  TWEN      2
#define  TWIE      0
#define TWAMR (*(volatile unsigned char *)0xBD)


/* USART0 */
#define UBRR0H (*(volatile unsigned char *)0xC5)
#define UBRR0L (*(volatile unsigned char *)0xC4)
#define UBRR0 (*(volatile unsigned int *)0xC4)
#define UCSR0C (*(volatile unsigned char *)0xC2)
#define  UMSEL01   7
#define  UMSEL00   6
#define  UPM01     5
#define  UPM00     4
#define  USBS0     3
#define  UCSZ01    2
#define  UCSZ00    1
#define  UCPOL0    0
#define UCSR0B (*(volatile unsigned char *)0xC1)
#define  RXCIE0    7
```

```
#define  TXCIE0    6
#define  UDRIE0    5
#define  RXEN0     4
#define  TXEN0     3
#define  UCSZ02    2
#define  RXB80     1
#define  TXB80     0
#define UCSR0A (*(volatile unsigned char *)0xC0)
#define  RXC0      7
#define  TXC0      6
#define  UDRE0     5
#define  FE0       4
#define  DOR0      3
#define  UPE0      2
#define  U2X0      1
#define  MPCM0     0
#define UDR0 (*(volatile unsigned char *)0xC6)

/* USART1 */
#define UBRR1H (*(volatile unsigned char *)0xCD)
#define UBRR1L (*(volatile unsigned char *)0xCC)
#define UBRR1 (*(volatile unsigned int *)0xCC)
#define UCSR1C (*(volatile unsigned char *)0xCA)
#define  UMSEL11   7
#define  UMSEL10   6
#define  UPM11     5
#define  UPM10     4
#define  USBS1     3
#define  UCSZ11    2
#define  UCSZ10    1
#define  UCPOL1    0
#define UCSR1B (*(volatile unsigned char *)0xC9)
#define  RXCIE1    7
#define  TXCIE1    6
#define  UDRIE1    5
#define  RXEN1     4
#define  TXEN1     3
#define  UCSZ12    2
#define  RXB81     1
```

```
#define   TXB81     0
#define UCSR1A (*(volatile unsigned char *)0xC8)
#define   RXC1      7
#define   TXC1      6
#define   UDRE1     5
#define   FE1       4
#define   DOR1      3
#define   UPE1      2
#define   U2X1      1
#define   MPCM1     0
#define UDR1 (*(volatile unsigned char *)0xCE)


/* bits */

/* Port A */
#define   PORTA7    7
#define   PORTA6    6
#define   PORTA5    5
#define   PORTA4    4
#define   PORTA3    3
#define   PORTA2    2
#define   PORTA1    1
#define   PORTA0    0
#define   PA7       7
#define   PA6       6
#define   PA5       5
#define   PA4       4
#define   PA3       3
#define   PA2       2
#define   PA1       1
#define   PA0       0
#define   DDA7      7
#define   DDA6      6
#define   DDA5      5
#define   DDA4      4
#define   DDA3      3
#define   DDA2      2
#define   DDA1      1
```

```
#define  DDA0     0
#define  PINA7    7
#define  PINA6    6
#define  PINA5    5
#define  PINA4    4
#define  PINA3    3
#define  PINA2    2
#define  PINA1    1
#define  PINA0    0

/* Port B */
#define  PORTB7   7
#define  PORTB6   6
#define  PORTB5   5
#define  PORTB4   4
#define  PORTB3   3
#define  PORTB2   2
#define  PORTB1   1
#define  PORTB0   0
#define  PB7      7
#define  PB6      6
#define  PB5      5
#define  PB4      4
#define  PB3      3
#define  PB2      2
#define  PB1      1
#define  PB0      0
#define  DDB7     7
#define  DDB6     6
#define  DDB5     5
#define  DDB4     4
#define  DDB3     3
#define  DDB2     2
#define  DDB1     1
#define  DDB0     0
#define  PINB7    7
#define  PINB6    6
#define  PINB5    5
#define  PINB4    4
```

```
#define  PINB3     3
#define  PINB2     2
#define  PINB1     1
#define  PINB0     0

/* Port C */
#define  PORTC7    7
#define  PORTC6    6
#define  PORTC5    5
#define  PORTC4    4
#define  PORTC3    3
#define  PORTC2    2
#define  PORTC1    1
#define  PORTC0    0
#define  PC7       7
#define  PC6       6
#define  PC5       5
#define  PC4       4
#define  PC3       3
#define  PC2       2
#define  PC1       1
#define  PC0       0
#define  DDC7      7
#define  DDC6      6
#define  DDC5      5
#define  DDC4      4
#define  DDC3      3
#define  DDC2      2
#define  DDC1      1
#define  DDC0      0
#define  PINC7     7
#define  PINC6     6
#define  PINC5     5
#define  PINC4     4
#define  PINC3     3
#define  PINC2     2
#define  PINC1     1
#define  PINC0     0
```

```
/* Port D */
#define   PORTD7   7
#define   PORTD6   6
#define   PORTD5   5
#define   PORTD4   4
#define   PORTD3   3
#define   PORTD2   2
#define   PORTD1   1
#define   PORTD0   0
#define   PD7      7
#define   PD6      6
#define   PD5      5
#define   PD4      4
#define   PD3      3
#define   PD2      2
#define   PD1      1
#define   PD0      0
#define   DDD7     7
#define   DDD6     6
#define   DDD5     5
#define   DDD4     4
#define   DDD3     3
#define   DDD2     2
#define   DDD1     1
#define   DDD0     0
#define   PIND7    7
#define   PIND6    6
#define   PIND5    5
#define   PIND4    4
#define   PIND3    3
#define   PIND2    2
#define   PIND1    1
#define   PIND0    0

/* PCMSK3 */
#define   PCINT31  7
#define   PCINT30  6
#define   PCINT29  5
#define   PCINT28  4
```

```
#define   PCINT27  3
#define   PCINT26  2
#define   PCINT25  1
#define   PCINT24  0
/* PCMSK2 */
#define   PCINT23  7
#define   PCINT22  6
#define   PCINT21  5
#define   PCINT20  4
#define   PCINT19  3
#define   PCINT18  2
#define   PCINT17  1
#define   PCINT16  0
/* PCMSK1 */
#define   PCINT15  7
#define   PCINT14  6
#define   PCINT13  5
#define   PCINT12  4
#define   PCINT11  3
#define   PCINT10  2
#define   PCINT9   1
#define   PCINT8   0
/* PCMSK0 */
#define   PCINT7   7
#define   PCINT6   6
#define   PCINT5   5
#define   PCINT4   4
#define   PCINT3   3
#define   PCINT2   2
#define   PCINT1   1
#define   PCINT0   0


/* Lock and Fuse Bits with LPM/SPM instructions */

/* lock bits */
#define   BLB12    5
#define   BLB11    4
#define   BLB02    3
```

```
#define  BLB01    2
#define  LB2      1
#define  LB1      0

/* fuses low bits */
#define  CKDIV8   7
#define  CKOUT    6
#define  SUT1     5
#define  SUT0     4
#define  CKSEL3   3
#define  CKSEL2   2
#define  CKSEL1   1
#define  CKSEL0   0

/* fuses high bits */
#define  OCDEN    7
#define  JTAGEN   6
#define  SPIEN    5
#define  WDTON    4
#define  EESAVE   3
#define  BOOTSZ1  2
#define  BOOTSZ0  1
#define  BOOTRST  0

/* extended fuses */
#define  BODLEVEL2 2
#define  BODLEVEL1 1
#define  BODLEVEL0 0


/* Interrupt Vector Numbers */

#define iv_RESET       1
#define iv_INT0        2
#define iv_EXT_INT0    2
#define iv_INT1        3
#define iv_EXT_INT1    3
#define iv_INT2        4
#define iv_EXT_INT2    4
```

```
#define iv_PCINT0          5
#define iv_PCINT1          6
#define iv_PCINT2          7
#define iv_PCINT3          8
#define iv_WDT             9
#define iv_TIMER2_COMPA   10
#define iv_TIMER2_COMPB   11
#define iv_TIMER2_OVF     12
#define iv_TIM2_COMPA     10
#define iv_TIM2_COMPB     11
#define iv_TIM2_OVF       12
#define iv_TIMER1_CAPT    13
#define iv_TIMER1_COMPA   14
#define iv_TIMER1_COMPB   15
#define iv_TIMER1_OVF     16
#define iv_TIM1_CAPT      13
#define iv_TIM1_COMPA     14
#define iv_TIM1_COMPB     15
#define iv_TIM1_OVF       16
#define iv_TIMER0_COMPA   17
#define iv_TIMER0_COMPB   18
#define iv_TIMER0_OVF     19
#define iv_TIM0_COMPA     17
#define iv_TIM0_COMPB     18
#define iv_TIM0_OVF       19
#define iv_SPI_STC        20
#define iv_USART0_RX      21
#define iv_USART0_RXC     21
#define iv_USART0_DRE     22
#define iv_USART0_UDRE    22
#define iv_USART0_TX      23
#define iv_USART0_TXC     23
#define iv_ANA_COMP       24
#define iv_ANALOG_COMP    24
#define iv_ADC            25
#define iv_EE_RDY         26
#define iv_EE_READY       26
#define iv_TWI            27
#define iv_TWSI           27
```

```
#define iv_SPM_RDY      28
#define iv_SPM_READY    28
#define iv_USART1_RX    29
#define iv_USART1_RXC   29
#define iv_USART1_DRE   30
#define iv_USART1_UDRE  30
#define iv_USART1_TX    31
#define iv_USART1_TXC   31

/* */

#endif
```

Author's Biography

STEVEN F. BARRETT

Steven F. Barrett, Ph.D., P.E., received the BS Electronic Engineering Technology from the University of Nebraska at Omaha in 1979, the M.E.E.E. from the University of Idaho at Moscow in 1986, and the Ph.D. from The University of Texas at Austin in 1993. He was formally an active duty faculty member at the United States Air Force Academy, Colorado and is now an Associate Professor of Electrical and Computer Engineering, University of Wyoming. He is a member of IEEE (senior) and Tau Beta Pi (chief faculty advisor). His research interests include digital and analog image processing, computer-assisted laser surgery, and embedded controller systems. He is a registered Professional Engineer in Wyoming and Colorado. He co-wrote with Dr. Daniel Pack six textbooks on microcontrollers and embedded systems. In 2004, Barrett was named "Wyoming Professor of the Year" by the Carnegie Foundation for the Advancement of Teaching and in 2008 was the recipient of the National Society of Professional Engineers (NSPE) Professional Engineers in Higher Education, Engineering Education Excellence Award.

Index

Printed in the United States
by Baker & Taylor Publisher Services